A Neurologist's
from the Steel Cage of Religion

A Long and Arduous
Personal Journey to Reality

Reynaldo Lazaro, M.D.

the Peppertree Press

www.peppertreepublishing.com

ISBN: 978-1-61493-828-6

Library of Congress: 2022910797

Printed: June 2022

"Blind belief in authority is the greatest enemy of truth"
  – *Albert Einstein (1879-1955), German-born theoretical physicist and developer of the theory of relativity*

"The idea that God is an oversized white male with a flowing beard, who sits in the sky and tallies the fall of every sparrow is ludicrous. But if by "God", one means the set of physical laws that govern the universe, then clearly there is such a God. This God is emotionally unsatisfying…it does not make sense to pray for the law of gravity."
  – *Carl Sagan (1934 -1996), astronomer, astrophysicist, astrobiologist, cosmologist, and planetary scientist*

"There is a fundamental difference between religion, which is based on authority, [and] science, which is based on observation and reason. Science will win because it works"
  – *Stephen Hawking (1942-2018), British theoretical physicist and cosmologist*

"The invisible and the non-existent look very much alike"
  – *Huang Po (850 AD), ancient Zen master and Chinese master of Tang Dynasty*

"When a person suffers from delusion, it is called insanity. When many people suffer from delusion, it is called religion"
  – *Robert M. Pirsig (1928-2017), American writer and philosopher*

"It is difficult to free fools from the chains they revere"
  – *Voltaire (1694-1778), French Enlightenment writer, historian and philosopher*

"The inhabitants of the earth are of two sorts: those with intelligence, but no religion, and those with religion, but no intelligence"
  – *Abu al- 'Ala al-Ma'arri (973-1057), Arab philosopher, poet, and writer*

DEDICATION

This book is dedicated to all truth seekers, free-thinking modern-day hominids, open-minded and farsighted mortals, and fellow escapees from the steel cage of religion.

TABLE OF CONTENTS

INTRODUCTION

"If you are religious at all, it is overwhelming probable that your religion is that of your parents. If you were born in Arkansas and you think Christianity is true and Islam false (knowing full well that you would think the opposite if you had been born in Afghanistan), you are the victim of childhood indoctrination"
– Richard Dawkins, author of "God Delusion"

"It could be ventured to understand obsessive compulsive neurosis as the pathologic counterpart of religious development, to define neurosis as an individual religiosity; to define religion as a universal obsessive compulsive neurosis
– Sigmund Freud (1856 – 1939), Austrian neurologist and founder of psychoanalysis

"Religion is poison because it asks us to give up our most precious faculty, which is that of reason, and to believe things without evidence. It then asks us to respect this, which it calls faith"
– Christopher Hitchens (1949-2011), British-American orator, journalist and columnist, and author of "God is Not Great: How Religion Poisons Everything"

Fifty-two long years have passed since my graduation from a Catholic medical school in Manila, Philippines, and I still find myself embroiled in never-ending controversies about religion and/or belief in supernatural being. However, I have remained entrenched in my scientific principles for very good reasons. I graduated without those coveted and sumptuously adorned Latin words (cum laude, magna cum laude, and summa cum laude) adjoining my name. I must add that I barely passed a "science" subject known as theology, which literally means "study of God" – or a study of an invisible being who, according to believers, created the billions of galaxies and trillions of stars, hundreds, thousands and millions of light years away from each other in this boundless universe – believe it or not – in just 6 days!!! Thus, apparently, I was simply a run-of-the-mill, average medical student, much to the chagrin of my parents, whose expectation of me was sky high, not to mention the tons of money they spent on my education. This, among many other factors, became the impetus that drove me to go full throttle to become head and shoulders above the rest in subsequent years as I began my postgraduate career in the land of a multicultural and multi-faith society known as the United States of America. Derek Jeter, a Yankee hall of famer, remarked, "It's not how you start the baseball season, it's how you end it." After all is said and done, he finished his career with five world series rings. However, it is inappropriate to compare myself to this great Yankee player, but I humbly say that I came 314 feet close (distance between the home plate and the right field in the Yankee stadium).

This book was written after the publication of several articles in various neurological, medical, and surgical journals over a span of 45 years, not to mention the diverse manuscripts I was asked to peer-review by the editorial boards of several scientific journals for publication. Overall, my scholastic records pale in comparison with those of other well-known neuroscientists and academicians, so I wanted to do something different by emplacing a "jewel in the crown," hence this tells-all book. It is the culmination of at least sixty years of personal observation of hypocrisies, contradictions, ironies and paradoxes that beset the practice of Catholic religion. Inexplicably and astonishingly, believers have turned a blind eye to these flustering and confusing aspects of this institution. Such is an unfortunate sequelae of childhood indoctrination but, luckily for me, my limbic lobe did not fall victim to such unjust and unscientifically blind process. My wings are free to spread and flap outside the steel cage of religion, and my mind is free to wander around and look at myriad of things in nature waiting to be explored by science – not by fallacious and delusive belief system.

It saddens me that my very religious parents did not live long enough to witness my efforts to make up for my subpar performance in medical school. My tenures as a research fellow in the Jerry Lewis Neuromuscular Disease Research Center at Vanderbilt University Medical Center in the Volunteer State of Tennessee and as an Associate Professor of Neurology and Co-director of the Muscular Dystrophy Clinic and Electromyography Laboratory at Albany Medical College in the Empire State of New York were the turning points in my career as an academician and a clinical neurologist— and now an unrepentant "escapee" from a social institution called religion.

Many years ago, when I joined an e-mail group organized by a fellow alumnus of my medical school, the exchange of ideas and experiences together with frequent reminiscing were daily and joyful events. Unfortunately, when I was tempted to express my candid and honest views on religion and God, the tone of some blogs and posts became sour, bitter, and too caustic for my classmates' lingual papilla and palate. As I write this book, my classmates' facial, glossopharyngeal and vagus nerves were reverberating wildly and depolarizing uncontrollably, ad nauseam. The temptation was inevitable when I found out that I was on a tiny and isolated island in the vast Pacific basin, which means I was, and still am, the only nonreligious (or areligious) student in the class of 1970. In that island, I do not look up to talk or listen to that invisible deity up in the sky supposedly present in all corners of the world who monitors everything billions of humans living on planet earth do every day, every hour, every minute, every second and every millisecond. Not surprisingly, many classmates whose literacy in religion is eerily excellent, expressed their disdain, disappointment, and antagonism. Acerbic and vitriolic attacks came pouring with some calling me a "different and strange" person for lack of a better word. One classmate from Peach State, who was treated previously and successfully for a hematological disorder, got pissed off (his own words) and never got in touch with me directly again. Another classmate criticized and "scolded" me, to put it mildly, within a week after reading my negative blog. Sadly, he died of complications from open-chest surgery several months later, and no further dialogue or debate took place. One classmate who lives in Bay State became a widow. Another classmate (a successful hematologist and oncologist) who is also a widow and lives in the Grand Canyon State, expressed her wholehearted and unequivocal support of my views about religion but she never got in touch with me again because of our political differences (any discussion on politics and religion can be divisive and toxic to friendship). Others, who were (and still are) my close friends in medical school, remained neutral and tight-lipped, but a week later, one of them (a successful psychiatrist practicing in the Blue Grass State) wrote, "My God, we're all brainwashed!" Some, I suspect, did not care. I then realized, neurologically speaking, that their cerebral neural network and synapses, those that stored the religious education they had acquired through the years in Catholic schools, stood still and frozen in time, or shall I say in neuropathological terms, "kindled, fossilized, and calcified." Now I understand, thanks to my education in neuroanatomy, neurochemistry, neurophysiology, and neuropathology, the education that most of my classmates did not have—and that's a major problem that sets us (and members of my family and relatives) several light years apart. Some fellow neurologists, I suspect, may have opinions and views similar to mine, but they prefer to shut off their speech center in the brain (known as Broca's area), zip their lips and deactivate their muscles of articulation for the sake of political correctness or for the fear of being ostracized and demonized.

Most of the information—or revelations—discussed in this book was obtained from numerous books and publications written by neurologists, psychologists, freethinkers, journalists, freelance writers, political and social activists, orators, columnists, politicians, archeologists, geologists, engineers, clergies, philosophers, biblical scholars, agnostics, and atheists, among many others. Being an open-minded and far-sighted person with expertise in the field of neurology, I find the works of these erudite scholars quite informative and enlightening. They changed my perspective about life and how it is influenced by molecular, gravitational and electromagnetic forces in this vast universe. Unsurprisingly, their works had been criticized caustically and spitefully by several religious pundits and erudite biblical scholars, who even labeled their research pseudoscience, fraudulent, and nonsensical. It is quite obvious that any valorous challenge to orthodox thinking will ignite traditional and conservative believers to react sternly. As the saying goes, "If the information is something new and innovative, and if it contradicts the accepted dogma, by all means, condemn it, suppress it, tear it and burn it down." The burning of Jewish books and the Talmud, which discusses the true identity of Jesus Christ, and the church's condemnation of Galileo for being anti-science and for holding the belief that the earth revolves around the sun, are few examples of inimical and pernicious reactions of believers when traditional views about their religion are challenged. Religion, per se, is a venerable institution but its devout followers can become fanatical and violent when their belief system is challenged. This behavior is mystical and commonplace among believers, and is the result of deep seated and firmly established belief in supernatural inculcated into their developing mind during childhood. Such behavior, which is discussed in Chapter 12, has neuroanatomical and neurochemical bases.

Christians, particularly Catholics who dare to read this book, will be interested in the presentation of a brief discussion on the prehistoric Philippines and the state of the country during the Spanish, American, and Japanese occupations, including the long-term effects of the colonization, atrocities, and abuses committed by the colonizers on the mental attitudes and personality of most Filipinos. Inconsistencies and controversies in the Bible, along with hypocrisies and contradictions in religion, are presented. The possibility of alien visitations and extraterrestrial influence on earthlings through millennia will excite the neurons of 21st century hominids. There are quotes about religion from prominent Americans, including various intellectuals that are interesting and thought-provoking (or maybe offensive to some). Quotes from Robert Ingersoll (1833–1899) and Thomas Paine (1737–1809) are particularly eye-opening. Ingersoll, the son of a Presbyterian minister, was a lawyer, writer, and orator. Paine was an English-born American political activist, philosopher, and political theorist. He was the author of "Age of Reason", a bestseller three-part book originally published in 1794, 1795, and 1807.

Neurological explanation for out-of-body experiences, together with relevant clinical literature on the subject, is presented. Complex issues and controversies resulting from mixing religion with politics and medicine will razzle-dazzle the thinking mind. The intriguing possibility that so-called "sins and deviant behavior" may actually be due to innate "dysfunctional" neural connections and abnormal activity of various neurotransmitters in the limbic system is also presented. If the believers elect to read this book, I suggest they buckle up because they may be offended (or intrigued); alternatively, they can simply read the discussion on Philippine history only and skip anything pertaining to religion and God. This book is for the open-minded and hyperopic individuals who wish to know the truth about the catholic religion and for those who wish to expand their neuronal synapses. It is a must read, especially for believers, and for them to realize the colossal void and light-year gap in their knowledge (this is not hyperbole) that badly needs to be filled with the wealth of information presented in this book. Yes, it's worth reading this eye-popping piece of work, regardless of your beliefs.

For those who wish to conduct a peaceful, intelligent, and professional discussion with me about this book – after reading it cover to cover – I strongly advise you to first read all the books, journal articles, periodicals, and online articles listed in the bibliography. I guarantee that you'll find the book thought-provoking, eye-opening, educational, irksome, agitating, worrisome, pestiferous, and entertaining. Books that are worthy of special mention are: The Bible Fraud by Tony Bushby, The Christ Conspiracy: The Greatest Story Ever Sold, by Acharya S, The Book Your Church Doesn't Want You To Read by Tim C. Leedom, Yahweh Encounters: Bible Astronauts, Ark Radiations and Temple Electronics by Ann M. Jones, The Genesis Race: Our Extraterrestial DNA and the True Origins of the Species by Will Hart, Flying Saucers and Science by Stanton Friedman, Chariots of the Gods by Erich von Daniken, Fingerprints of the Gods by Graham Hancock, and The Origin of Existence by Fred Adams. Various neuroscience articles, on-line sources and periodicals cited in this book provided startling, provocative and jaw-dropping information.

Am I proselytizing? No, I am not. I am simply expressing in writing the knowledge I have properly accumulated and carefully stored, and eventually "kindled" in the neural fabric of my limbic lobe through years of living in a room with a panoramic view of the world. If, after reading this book, you find me loathsome and despicable (or perhaps an instrument of the devil), be informed that you will only solidify my honest views about religion and God. However, if you feel that the book broadened your knowledge in all directions—east, west, north, and south—I honestly and proudly say, "Congratulations!!" So now you can use the book as a rescue tool to help you escape from the steel cage of religion. For those staunch and profoundly imbued believers in catholic faith who wish to remain inside the cage that they venerate for many years since childhood (before they reached the age of reason), may the light of revelation brighten your room, stimulate and invigorate your cerebral synapses and, hopefully, untangle the tortuous and hardened neural fabric in your limbic lobe, someday – if tomorrow comes.

ACKNOWLEDGMENT

I give credit to my roommate of 52 years, Zeny, whose untiring support, understanding, and inspiration made me an established academician and a solid clinical neurologist, and to my children (all first-generation Filipino-Americans) and to Scribendi Editing Service in Ontario, Canada, for editing the book. Zeny is a devout believer, and I am not, but Cupid has done a splendid job of keeping our hearts (and neurologically, our brain) tied together. My youngest, Maria, provided interesting information on the subject of the prehistoric Philippines. Special thanks go to my classmate, Dr. Alfredo I. Custodio, the author of the book titled "You've Got M@il" and a successful practicing surgeon in the Show Me State of Missouri, whose encouragement and support were invaluable. He is the chief of the Grammar Police Force of our e-mail group, and he did his job effectively. He apprehended and handcuffed many classmates (I mean, he did it clandestinely together with my naughty participation) and sent them back to grammar school. He is a loyal fan of the St. Louis Cardinals baseball team that won 11 world series rings (!!!!!!!!!!!!), but my Yankees have 27 (!!!!!!!!!!!!!!!!!!!!!!!!!!!!!). I affectionately call him "Uncle Dave" because of the uncanny similarities between his style of writing and that of Mr. Dave Barry, a Pulitzer-winning humor columnist for Miami Herald. My pertinaciously religious classmates' uncomplimentary remarks in past years were provocative but at the same time inspirational and strongly motivational. Some of my loyal and erudite patients, who grew up in very religious families but luckily and auspiciously managed to free themselves from the powerful influence of religious obsessive-compulsive practices, provided fuel to the driving force that compelled me to write this book.

Thank you all (or y'all for those who live down south), and may the God (or gods) of your religion bless you. Pax vobiscum.

CHAPTER 1

The Arrival of Ferdinand Magellan and his army of Christians with the Bible in one hand and a sword in the other, and prehistoric Philippines

> "The pioneers and missionaries of religion have been the real cause of more trouble and war than all other classes of mankind"
> – *Edgar Allan Poe (1809-1849), writer*
>
> "They came with a Bible and their religion – stole our land, crushed our spirit ... and now tell us we should be thankful to the "Lord" for being saved"
> – *Chief Pontiac (1720-1769), Ottawa Indian chief*

On March 16, 1521, Ferdinand Magellan, a Portuguese explorer who sailed under the Spanish flag, set foot on the soil of Homonhon Island in eastern Samar. He named it Isla San Lazaro and claimed it for Spain. Upon seeing more islands in the vicinity, he named them Filipinas, in honor of King Philip II of Spain. Two weeks later, after erecting a cross on the island, the first Catholic mass was held in Limasawa, an island in Leyte, conducted by a Spanish Friar, Fr. Pedro Valderama, in the presence of two rajas—Rahaj Ciagu and Rajah Calambu—who were subsequently baptized along with several natives. And so began the process of Christianization and religious education (a.k.a. indoctrination) in this archipelago of 7000 islands with some 120 to 187 languages. In 1543, after the archipelago became officially named Las Islas Filipinas by Ruy Lopez Villalobos, a

Spanish explorer, the Philippines became part of the Spanish empire that included Mexico.

How two groups of people from different parts of the world were able to communicate with each other, with one convincing the other to adopt a religion based on a belief system, was difficult to fathom, but it happened. However, in this world, with an estimated 4000 religions, divisiveness, discrimination, conflicts, bigotry, and antagonism are inherently normal human behavior. Magellan suffered from the consequences of religious and social differences when a local chieftain in Mactan, by the name of Lapu-Lapu, refused to join the band wagon of Christianized Filipinos and revolted. In the local battle that ensued, Magellan was killed, marking the first documented Filipino resistance against foreign occupiers.

Obviously, unbeknownst to Magellan was the long history of the Philippines, which began at least 709,000 years ago during the Pleistocene geological epoch, often referred to as the Ice Age, which began around 2.6 million years ago and ended around 11,700 years ago. Pleistocene stone tools and remains of butchered Rhinoceros philippinensis associated with hominin activity were discovered in 2018. A species of archaic humans, Homo luzonensis, was present on the island of Luzon approximately 67,000 years ago, while the earliest known modern human was known to exist in the caves of Palawan about 47,000 years ago. The prehistoric Philippines was later inhabited by the Negritos scattered all over the major islands and are believed to be descendants of groups of Homo sapiens that migrated to the Philippines during the upper Pleistocene era from Southeast Asia mainland, but not from Africa. Today, the mongoloid people of the Philippines outnumber the Negritos by 4000 to one. Around 3000 B.C., seafaring Austronesians migrated southward from the island of Formosa (Portuguese word for "beautiful"), now known as Taiwan. Following settlements in the archipelago, various economic and political systems, along with social organizations and stratifications, developed. These societal groups had their own religion influenced by Hindus, Buddhists, Islam from Arabia, and China. To add credence to the existence of these societal and cultural groups is the discovery of the Laguna Copperplate Inscription, measuring 20 x 20 cm, in 1989. The plate was found by a laborer during a dredging activities in Lumbang River near Laguna de Bay. The plate was later offered for sale to the National Museum of Manila, and translated by Antoon Postma, a Dutch Anthropologist. The inscription was similar to the ancient Indonesian script of Kawi dated 822 AD Saka year, an old Hindu Calendar date which corresponds to the year 900, about the same time as the official Chinese Song dynasty History of Song in 972. The text is old Malay with some words derived from Sanskrit, Old Javanese and old Tagalog.

Most historians use the term "discovery" to describe the arrival of an explorer on a certain island. The arrival of Magellan on the island of Samar, Philippines, was one example. The term seems inappropriate since the island and several islands up north and down south were already inhabited by people of various ethnic groups with different religious beliefs. In all fairness, however, he did not have a chance to explore the numerous northern and southern islands, each inhabited by people with various socioeconomic, political, and religious affiliations and practices. By the same token, the claim by historians that Christopher Columbus discovered America should be reconsidered, as discussed in the fascinating book written by Peter Schrag and Xaviant Haze titled, "The Suppressed History of America." In this book, they discuss the discoveries of Meriwether Lewis (1774–1809), an American explorer, public administrator, and politician. Lewis revealed several evidences that America was inhabited by a highly evolved ancient civilization that might have been descendants or had made contacts with explorers from Europe and Asia, long before the arrival of Columbus. Members of this civilization were called the Olmecs, the earliest-known Mesoamerican civilization, which was prominent between 1200 and 400 B.C.E. They built large settlements, established trade routes, and developed religious rituals. The culture, like the Mayan culture, eventually became extinct for no clear reason, but they left behind giant statues now known as Olmec heads, found in the jungles of Mexico, Guatemala, Honduras, Costa Rica, El Salvador, and Belize.

Back in the pre-colonial period, animism was widely practiced, but today, thanks to Spanish influence, only a few indigenous tribes practice the religion. What is Animism? Ancient inhabitants believed that gods and spirits guarded rivers, forests, and mountains. It is the belief in spiritual beings and that the world is inhabited by spirits and supernatural entities, both good and bad. Polytheism was also practiced together with beliefs in immortal gods and goddesses, deities, and various divine beings that possessed supernatural powers akin to those described in Greek mythology and those described in ancient Egyptian, Sumerian, Indian, Mayan, and Chinese literature. Such religious practices were unacceptable to Magellan and his crew, who circumnavigated the world to promote trade, search for valuable spices, and spread Christianity along with belief in a god that exists in three forms!! However, his effort to inculcate Christian principles in the minds of unsophisticated and some non-gullible natives was met with stiff resistance—which eventually led to his demise. Differences in religious principles and practices, together with bigotry and elitism, remain pervasive in world society and have always been the root of endless wars among Christians and non-Christians since the beginning of time.

There were similarities between Magellan and Columbus. Both travelled west from Europe to find a route to Asia and to promote trade. The former reached the Philippines, while the latter reached North and South America. Both spread Christianity, but at the same time carried racist prejudices toward the natives. Magellan reportedly bragged that their armor and weapons were far superior to the natives' spears and bolos, but the natives outnumbered the Spaniards, and in the ensuing war, he was killed by Lapu-Lapu, a local chieftain. Native Americans were perceived as "heathens," and within a century, after the so-called discovery of America, millions of natives were killed by Christian explorers, all in the name of Christianity, a religion based on Jesus Christ, the son of God, who preached peace to mankind. Go figure that out.

The general perception among Filipinos is that if not for the Spaniards, the Philippines would have been an Islamic country, which seems to imply that Christianity is their preferred and perhaps the better religion. This is fallacious and misleading. This premise cannot be accepted by almost two billion Muslims living in this world. Moreover, accepting a social or religious institution like Christianity, whose foundation is based on beliefs and hearsays propagated by unscientific and unsophisticated people without media scrutiny and technological documentation, is absurd. The annihilation of thousands of lives of innocent natives who were living peacefully and minding their own business prior to the arrival of those explorers, and all in the name of Christian religion, is beyond human reasoning.

Has anyone wondered who gave them (or their predecessors) information about the existence of civilizations west of Europe and across the Pacific? Was the reported sighting of UFOs by Columbus hours prior to setting foot on the soil of America relevant or a hoax?

Praise the Lord for helping Magellan and Columbus spread Christianity outside of Europe. ⬠

CHAPTER 2

Over 300 years of Spanish colonialism, reorganization of Philippine society, and various abuses committed by Spaniards toward Filipinos

> "Anyone who knows history will recognize that the domination of education or of government by any one particular religious faith is never a happy arrangement for the people
> – *Eleanor Roosevelt (1884-1962), in a letter to Cardinal Spellman, July 23, 1949.*
>
> "If Jesus were alive today, the last thing he'd be is a Christian"
> – *Mark Twain*
>
> "If Jesus had been killed twenty years ago, Catholic school children would be wearing little electric chairs around their necks instead of crosses"
> – *Lenny Bruce (1925-1966), American comedian, activist and social critic*

Upon the order of King Philip II, Luis de Velasco, the viceroy of New Spain (now Mexico), commissioned Miguel Lopez de Legazpi to undertake an expedition to the Philippines with the intent of colonizing it. In April 1565, he reached Cebu, established the first Spanish settlement, and became the first governor of the Philippines. Six years later, the city of Manila was established and became the capital of the new colony and a major trading port in East Asia.

Although he encountered very strong resistance from the Muslims in the southern Philippines (the Mindanao, Jolo, and Sulu archipelago), he and his chaplain, Andres de Urdaneta, successfully converted most Filipinos living on the major island of Luzon, where Islamic influence was weak, to Christianity. The diverse religious practices became homogeneously Christian. This paved the way for various changes in the political, socioeconomic, and educational systems in the country. Infrastructures were built, and on April 28, 1611, the Pontifical and Royal University of Santo Tomas was founded by a Spanish Friar, Miguel de Benavides. The Augustinians and the Franciscans brought literacy to the people. Intermarriages between native inhabitants and the Spaniards gave birth to "mestizos," a name given to mixed genetic ancestry (some were children of sexual assaults committed by Spanish priests toward Filipino women). Filipino-Chinese mestizos are the largest mestizo population, while Filipino-Spanish mestizos are fewer but prestigious, socially significant, and influential minorities.

What seemed to be honorable and great accomplishments of the Spaniards toward modernizing their colony were overshadowed and badly tainted by the abuses committed by the clergies, government elites, and Spanish soldiers. Filipinos were treated as second-class people, were taxed heavily, were forced to engage in hard labor (building bridges, churches, and roads), and were punished or killed if they disobeyed. A historical legacy was sexual misconduct committed by priests toward Filipino women and, along with mistreatments of the Filipinos, led to the call for independence by the new class of Madrid- and Barcelona-educated Filipinos, led by Jose Rizal, the national hero of the Philippines and the foremost symbol of Filipino nationalism. Exposure to such abuses is depicted in the two books he wrote in Spanish (Noli Me Tangere and El Filibusterismo). Other Filipino intellectuals who studied in Spain, such as Graciano Lopez Jaena, Marcelo del Pilar, Mariano Ponce, and Antonio Luna, also called for their country's independence. On February 17, 1872, three martyred priests, Mariano Gomez, Jose Burgos, and Jacinto Zamora, were executed on charges of subversion and alleged complicity in the uprising of workers in the Cavite Naval Yard. Twenty-four years later, on December 30, 1896, Jose Rizal was executed for sedition.

Although the incidence of cases of pedophilia and decades of predatory sexual assaults committed by priests during the colonial period is unknown, it will not surprise anyone in these modern days of media attention and scrutiny that such perverted and deviant sexual behavior might have occurred but intentionally buried in oblivion with threats of severe repercussion against victims by higher authorities. Today, various law firms are soliciting cases of sexual abuse committed by clergies. This is a worldwide problem, and several dioceses have filed for bankruptcies or settled sexual abuse cases. Over the last 20 years, the Catholic church in the Philippines apologized for sexual abuses committed by 200 priests, including sexual harassment of women by several priests in 2003, a 17-year-old female sexually abused by a priest, and an attempted case of sex with a 13-year-old girl by a young priest. In 2018, an American priest was arrested for sexually molesting dozens of Filipino young boys over a period of decades.

Why in the world would Filipinos adopt a religion that was forcibly shoved into their throat by foreign occupiers who declared themselves children of God, and that instilled into their mind that Catholicism is "The Religion" and nothing else is beyond rational comprehension—but it unfortunately happened. Now, the Philippines is the only Christian nation in Asia where at least 86% of the population is Roman Catholic. They are very religious people, rich and poor alike. The country, however, with a population of 111,600,000, is still struggling with a relatively high poverty rate, especially in rural areas. Some people ask why, despite the prayers, a substantial number of Filipinos still live below the poverty level. The reasons are likely multifactorial and are alluded to in other chapters of this book. Natural disasters, particularly typhoons and earthquakes, are significant factors. Overpopulation is clearly a major factor, and let's figure this out. The state of Arizona has a land area of 113,990 square miles (Philippines' land area is 115,831 square miles) and has a population of 7.2 million as of 2021. The US is 33 times bigger than the Philippines and has a population of 300 million. Economically, it appears that non-Catholic countries such as Japan, South Korea, Taiwan, and China are prosperous and are able to compete with world economies, particularly with reference to automobile and technological equipment exports.

God, please heal the scrupulosity of downtrodden Filipinos. ℘

CHAPTER 3

The British occupation, the defeat of the Spanish empire, and the dark side of American colonialism

"Our nation was born in genocide when it embraced the doctrine that the original American, the Indian, was an inferior race. Even before there were large number of Negroes on our shore, the scar of racial hatred had already disfigured colonial society. From the 16th century forward, blood flowed in battles over racial supremacy. We are perhaps the only nation which tried as a matter of national policy to wipe out its indigenous population. Moreover, we elevated that tragic experience into a noble crusade. Indeed, even today we have not permitted ourselves to reject or feel remorse for the shameful episode. Our literature, our films, our drama, our folklore all exalt it. Our children are still taught to respect the violence which reduced a red-skinned people of an earlier culture into a few fragmented groups herded into impoverished reservations."
– *Dr. Martin Luther King Jr (1929 – 1968), American Baptist minister and leader of American Civil rights movement.*

"I hate imperialism, I detest colonialism. And I fear the consequences of their last bitter struggle for life. We are determined, that our nation, and the world as a whole, shall not be the play thing of one small corner of the world"
– *Sukarno (1901-1970), first president of Indonesia*

The occupation of Manila by the Kingdom of Great Britain for 20 months, that is, from 1762 to 1764, was relatively inconsequential in terms of religious influence. It was an offshoot of the Seven Years' War between Britain and France, during which time Spain became involved when it sided with France. The British wanted to use Manila as a trading port with China, but they faced stiff resistance when Filipino troops supported the colonial government, enabling them to limit their occupation to Manila and Cavite. Following the end of the war and the signing of the Treaty of Paris, the British left Manila and Cavite in April 1764.

On April 21, 1898, a war broke out between the US and Spain. The causes were multifactorial but had been brewing due to the ongoing struggle of Cuba against Spanish rule, together with the mysterious explosion of the U.S.S. Maine in Havana Harbor. Economic reasons and the strategic location of the Philippines were also significant. The desire to spread Christianity was also in the mix, even though Spaniards had already accomplished it during the colonial period. The US Navy, under the command of George Dewey, defeated the Spanish fleet on May 1, 1898, marking the beginning of the 50-year American presence in the Philippines.

Filipino revolutionary leaders, led by Emilio Aguinaldo, went in exile to Hong Kong after failing to destabilize the Spanish rule. Much to his chagrin, upon his return to the Philippines, with the help of Americans, the independence he declared in June 1898 did not materialize. President McKinley declared that Filipinos were incapable of self-government. Most Americans also believed that Anglo-Saxons were far superior to Filipinos in many ways. The Filipino-American war erupted when US troops fired on Philippine troops in February 1899. The war lasted from February 4, 1899 to July 2, 1902, and killed over 4200 Americans and over 20,000 Filipinos, including a young general, Gregorio del Pilar, the hero of the Battle of Tirad Pass. Filipino fatalities were probably higher and, according to another source (A People's History of the United States, 1980),

300,000 Filipinos were killed in Batangas alone. Another source (William Pomeroy's American Neocolonialism, 1970) cites 600,000 Filipinos killed in Luzon by 1902. It was obvious that Filipinos had no chance against the superior weaponry of Americans.

During the American occupation, it became obvious that the Americans grossly lied about giving independence to the Filipinos and that their occupation of the country was simply self-serving and egotistical. The war crimes against Filipinos were indescribable. They exploited and vilified Filipino heroes. Americans who came to the Philippines at the end of 19th century called the Filipinos, "brown, short and tailless monkeys"—derisive remarks made by citizens of a Christian nation to citizens of another Christian nation. Go figure.

On October 15, 1900, Mark Twain wrote in his New York Herald column, "I have read carefully the Treaty of Paris, and I have seen that we do not intend to free but to subjugate the people of the Philippines. We have gone there to conquer, not to redeem. It should, it seems to me, be our pleasure and duty to make those people free, and let them deal with their own domestic questions in their own way. And so I am an anti-imperialistic. I am opposed to having the eagle put its talons on any other land."

Stratification of race and affiliation with certain religion influence the way humans regard each other. In the US, the relationship between race and religion has always been complex, controversial and fragile. This relationship can break down when the innate and subconscious racial prejudice that some people harbor in their central nervous system become overtly expressed. Anti-Filipino racial politics and xenophobia in the US grew during the early 1930s. A judge in California referred to Filipinos as "scarcely more than savages." The North Monterey Chamber of Commerce referred to them as "ten years removed from a bolo and breechclout." A newspaper report quotes a white racist terrorist referring to Filipinos as "goo-goos." Many Americans thought that Filipinos were uncivilized and wild people. Many were reported to be mistaken for Chinese or Japanese.

In one photograph of a restroom in Stockton, there was a sign on the door that read, "POSITIVELY NO FILIPINOS ALLOWED.

Praise the Lord, and thank you, Jesus. ൚

CHAPTER 4

The Japanese occupation of the Philippines and the unmasking of the effects of years of oppression, humiliation, and subjugation of traits on Filipino ways of life.

> "The enemy is in front of us, the enemy is behind us, the enemy is to the right and to the left of us. They can't get away this time."
> – *General Douglas MacArthur (1880-1964), American military leader and Field Marshal to the Philippine Army during WW II*

> "Our actions brought suffering to the people in Asian countries. We must avert our eyes from that."
> – *Shinzo Abe, address to joint meeting of the U.S. congress, April 29, 2015*

Ten hours after the Japanese attacked Pearl Harbor on December 7, 1941, the invasion of the Philippines and World War II began. General Douglas MacArthur retreated to Australia on March 11, 1942. Despite all the prayers they learned during Spanish colonization, several thousand Filipinos died from Japanese atrocities. Several women were raped. It was a massacre infamously called "The Rape of Manila." The occupation lasted for over three years, but the Filipinos' effective guerilla resistance enabled MacArthur to return to the Philippines and save the country. Aware that they would be defeated by the Americans, the Japanese military conducted suicidal missions on the ground and on the sea, to no avail. Half a million Filipinos died during the Japanese occupation.

Could it be that the oppression, humiliation, persecution, and insults suffered by Filipinos in the hands of Spaniards, Americans, and the Japanese groomed the general personality of the majority of Filipinos? They became the "punching bags" of imperialistic and opportunistic foreign occupiers. Most Filipinos are, in general, shy and inhibited, less assertive, courteous, obedient, and humble. They are resilient and adaptable. Some of them are overly superstitious. The "bahala na" attitude, which expresses courage and faith in God, still pervades the country. It is a form of invocation or silent prayer that leaves everything in God's hands. Rebellious tendency is reflected in the last few words of their national anthem, "ang mamatay ng dahil sa iyo"—which means "willingness to die because of you" or "to die for my country." Perhaps the tone of the national anthem would have been different, and not bellicose, if not for the repeated oppressions suffered under the hands of the occupiers.

Rewards obtained from prayers are distributed unfairly. Overall, the religiosity of a great majority of Filipinos is not rewarded by economic prosperity or good health. Poor people pray every day, but they remain jammed and trapped in slum areas and squatter settlements, while the rich who own several rental properties and live in houses surrounded by concrete fences also pray regularly but are rewarded obscenely. The children of the rich, especially descendants of Spanish elites and mestizos, attend private parochial schools. There is clearly a mile-wide gap between poverty and affluence in the country called "Pearl of the Orient Seas." Infectious diseases are prevalent. Street crimes, as well as graft and corruption among politicians, are endemic. Hypocrisies and elitism, especially among the affluent, are commonplace—and yet they pride themselves for being "religious," with a strong belief in the God that they believe, which inexplicably exists in three forms. They feel "relieved and refreshed" after spending a few minutes talking to a priest, who may incidentally be a pedophile or a sexual abuser, inside the confession box, only to commit the same offenses days later. The awesome sight of the great Filipino boxer, Manny Pacquiao, who after smashing the head of his opponents and knocking them out cold would walk towards the neutral corner of the ring after the fight, has now turned into appalling sight of him kneeling and praying. He recently retired and is planning to run for presidency, but I hope that this great fighter will not get infected with a chronic infectious disease known as "graftiodes corruptionensis," a highly contagious disease endemic in the Philippines. I should add that politicians who suffer from this disease go to church every Sunday and routinely make the sign of a cross when passing by a church.

What kind of people would have Filipinos been if not for at least 350 years of colonization and subjugation of their traits by the Spaniards (300 years), Americans (48 years), and Japanese (3 years)? What if the Philippines was never under any foreign rule in all those years? I venture to say that, if not for those colonizers, the religious, scientific, social, political, and economic aspects of their lives would have been different. Theoretically, given the vast resources in this country, the economy and many aspects of their lives, without hindrance from population overgrowth and superannuated and antiquated religious practices, would have been topnotch in Asia. Filipinos can only look back, frustratingly and painstakingly.

God, why did you allow those colonizers to oppress the Filipinos for hundreds of years? What did they do wrong? Glory to You in the highest. ✑

CHAPTER 5

The religious indoctrination of the innocent and developing mind

"Anyone who can worship a trinity and insist that his religion is monotheism can believe anything."
– *Robert A. Heinlein*

"If religious instructions were not allowed until the child has attained the age of reason, we would be living in a quite different world"
– *Christopher Hitchens*

"Finding that no religion is based on facts and cannot be true, I began to reflect what must be the condition of mankind trained from infancy to believe in error"
– *Robert Owen (1771–1858), a Welsh textile manufacturer, philanthropist, and social reformer.*

"The family that prays together — is brainwashing their children"
– *Albert Einstein*

"Let children learn about different faiths, let them notice their incompatibility, and let them draw their own conclusions about the consequences of that incompatibility. As for whether they are "valid," let them make up their own minds when they are old enough to do so."
– *Richard Dawkins*

Children born to Christian or Muslim parents are called "Christian or Muslim" children, depending on the tutelage of their parents or clergies. Realistically, these children, as soon as they come out of their mother's womb, do not have religious preference, ideology, prejudice, or bigotry implanted in their brain, plain and simple. They do not enter the world with preconceived or independent thoughts about God or religion. A set of twins, raised by two parents, one a Catholic and the other a Muslim, will certainly develop contrasting religious practices as they grow and are taught to worship and pray. They will probably hate or even kill each other. Children raised by atheist parents who are trained to recognize right from wrong will grow up wondering why Christian, Jewish, and Muslim friends go to a place of worship and worship an invisible god.

The relatively meager neuronal axo-dendritic synapses in children's brain make indoctrination and the "kindling" process relatively easy, and when repeated on a regular basis, the information stored becomes hardened and fixed. Such a process is the physio-biochemical and anatomical basis of indoctrination in religion and of the learning system in a science institution—but with one clear difference—the former is based on belief, while the latter is based on facts that can be perceived by special senses.

Despite the numerous contradictions that beset religion, believers always turn a blind eye to this issue. Being religious is taken for granted. If a child begins to question the existence of God, parents would always convince the child to be on the "safe" side to guarantee a place in heaven. This is a powerful message inculcated into the young mind, so powerful that it becomes almost always irreversible. However, many believers refuse to admit that they were indoctrinated during childhood. Although teaching morality to children must be standard in all curricula, I find it disconcerting that it has to be contingent or amalgamated with God and religion. It is noteworthy that professionals, such as most medical doctors, will not feel offended when a colleague presents an opinion that contradicts their diagnosis or the results of a research study. This is not the case in religion. Regardless of the large amount of scientific knowledge imbued into the brain of some professionals, specifically medical doctors, when someone challenges one's faith, that person is likely to be chastised and ostracized.

Four words—God, heaven, hell, and sin—are constantly and repetitively instilled and nurtured inside a child's mind by the elderly and clergies. Sin and eternal punishment in hell, along with belief in God, absolution, forgiveness, and the reward of going to heaven, are always incorporated in all prayers. In most households, praying before going to bed or before eating dinner is standard or a must among very religious families. Missing Sunday mass is considered a sin, despite being taught that God is omniscient, omnipotent, and omnipresent. In the Philippines and other former colonies of Spain, religious processions that involve praying while walking and carrying statues of Jesus Christ and saints are commonplace. To devout Christians, attendance in the procession is uplifting and inspiring. All these are compulsive behavior performed to allay and neutralize anxiety, fear and guilt – all unpleasant feelings induced by the process of indoctrination.

Idolatry is not allowed or condoned in Islam but is the opposite in Christianity, particularly in Catholicism. For Catholics, it is common practice to move up to the altar, where statues of saints are securely placed, and to kiss their apparel vestments after the conclusion of the mass. I should add, I have seen many parishioners slip and fall in the altar and sustain multiple contusions in their extremities and face. Making the sign of a cross with wet middle fingers after dipping them in "holy" but very likely contaminated water and doing handshakes to wish parishioners "peace be with you" were routine practices before the COVID-19 pandemic. Years ago, before the pandemic, my wife and I had experienced a somewhat embarrassing and uncomfortable situation when one parishioner sitting behind us politely refused to shake hands with us to wish her peace and goodwill, while those sitting in front of us immediately applied hand sanitizer after shaking hands with us. Their actions were medically understandable, but we obviously felt offended. May the Lord bless those people.

Despite the natural androgen hormonal surge during the pubertal and adolescent years, onanism is considered a sin by Catholics. Spewing bad words, lying, and cheating are sins. Depending on the "severity" and number of sins, confessors are given penance by reciting prayers many times to be forgiven. Chastity belts are sometimes worn per order of the priest to avoid sexual intercourse or masturbation.

During my teens, I recall that one of my sisters and a stepbrother attended a De Colores congregational camp separately for one week to "wash" their soul. This congregational meeting is still a common practice in former Spanish colonial countries. The attendees are called "Cursillistas" (which means those who attended a short course in Christianity). During this period, the attendees gather together, pray several times a day, share their feelings of guilt, and reveal their sins, particularly the ones they committed against their friends and relatives. When my sister, and later my stepbrother, arrived home after a week of grueling and emotionally draining daily prayers and confessions, they were greeted by former attendees, who sang various religious songs while loudly clapping their hands. They entered the house walking on their knees, crying and sobbing and expressing their remorse for the sins they committed. I was baffled by this strange scene I witnessed. Several weeks later, they both tried to convince me to attend the congregation because it was their impression that I was riddled with sins and that by attending the congregation was a good way to wash the soul—but my bewildered mind strongly rejected their recommendation. Few months later, I found out that my stepbrother remained a womanizer, wife abuser, and gambler, and he was also involved in graft and corruption cases while working for the government. My sister, I found out years later, made unfair and clandestine sale of our family inheritance properties to benefit herself and her family members without my prior knowledge and consent. Obviously, it was arduous to pursue matters legally because of the vast ocean that separates the Philippines and the US, so I remained meek and with subdued resentment. These pendular movements from the sinless to a sinful state among some Filipinos are appalling and seemingly casual social phenomena, regardless of religious affiliation. From the scientific standpoint, fragility and instability of human behavior are realistically physiologically and biochemically mediated, but to believers, everything is spiritual and can be rectified by prayers and confession. Hallelujah, praise the Lord. I am forgiven!!

The practice of walking on one's knees from the church entrance to the altar is common in the Philippines, and it always reminds me of the wife of a former president of the Philippines, who was witnessed several times walking to the altar on her knees while praying using a diamond holy rosary (each bead was at least one carat). The former president and his wife and their cronies pocketed millions of dollars given to Philippines by the US government as part of the economic aid. The ill-gotten wealth was used to fund their lavish lifestyle. I should add that behind and adjacent to the church was a prostitution house where my religious late cousin was a frequent visitor. Go figure.

During the Lenten season, I witnessed interesting religious practices among my relatives and friends living in the Philippines and California. They travelled to and prayed in several churches within a 50-mile radius all day long on Good Friday. They believe that such practices will absolve them of their sins. I later learned that some of them, including their parents, entered the U.S. illegally and obtained disability benefits fraudulently.

Bloodcurdling and gruesome religious festivals that take place on Good Friday are worthy of special mention in this deeply religious nation of the Philippines. Dozens of compunctious devotees re-enact Christ's sufferings by having their palms and feet nailed on the cross by participants dressed in Roman centurions, after carrying the cross under the heat of the sun or after self-flagellation using wooden whips. Devotees believe that such sufferings would wash away their sins. The tradition, which takes place in central Luzon, attracts onlookers and tourists every year. It is condemned and discouraged by Catholic leaders and health care professionals. Such practice clearly fits the definition of fanaticism, an unfortunate sequelae of religious indoctrination in which no amount of re-education can untangle the cemented neural fabric of the limbic lobe of those admirable but unsophisticated and gullible religious devotees. Obrigado (thank you), Fernao de Magalhaes (Ferdinand Magellan)!

Holy Mary, Mother of God, pray for us sinners now and at the hour of our death, Amen. ❧

CHAPTER 6

Learning the science of medicine in parochial schools

"The church has had the field for eighteen hundred years. For most of this time it has held the sword and purse of the world. For many centuries it controlled colleges and universities and schools. It had within its gift wealth and honor. It held the keys, so far as this world is concerned, of heaven and hell - that is to say, of prosperity and misfortune. It pursued its enemies even to the grave. It reddened the scaffold with best blood, and kept the sword of prosecution wet for many centuries. Thousands and thousands have died in its dungeons. Millions of reputations have been blasted by slanders. It has made millions of widows and orphans, and it has not only ruled this world, but it has pretended to hold the keys of eternity, and under this pretense it has sentenced countless millions to eternal flames."
– *Robert G. Ingersoll*

"The word God is for me nothing more than the expression and product of human weaknesses, the Bible a collection of primitive legends which are nevertheless pretty childish. No interpretation no matter how subtle can change this."
– *Albert Einstein*

"Theology is ignorance with wings"
– *Sam Harris*

Most agree that science and religion are not miscible; the former is based on observation, experimentation, reasons, and facts that can be measured, while the latter is based on beliefs, and nothing more. It is therefore ironic that some students who later become scientists learn the art of medicine from schools run by clergies. In those institutions, it is routine (and required) to pray before each class begins. Theology, the study of God, is part of the academic curriculum, but to some critics, it seems absurd, to say the least, to study a being that one cannot see or has not been seen. Believers would argue that the evidence is all recorded in the Bible and in the various accounts of the miracles of Christ "documented" 2000 years ago. Unfortunately, during that time, there was no media scrutiny and no scientists to witness or authenticate so-called supernatural events. How the idea of God originated or evolved is another issue, but it is apparent that thousands of years of proselytizing by clergies have induced a "kindling" effect on the neural loop that interconnects the orbital and frontal cortex, cingulate gyrus, striatum, and thalamus of religious people. Sad, but true.

During my premedical education, the process of learning theology was a nerve-wrecking and anxiety-generating experience. Some priests used dramatic methods of proselytizing. They would raise their voice or even shout at the top of their lungs, enough to shutter the students' tympanic membrane, to make a point. It was sensational or even theatrical, but nobody dared to make side comments after the teaching session ended. Such is the power of indoctrination, which can easily groom and mold the minds of the oppressed and downtrodden people like the Filipinos.

Politics, cheating, and corruption are endemic, even in parochial schools. There was an occasion when I was forced to repeat a course in Reserve Officers' Training Corps (ROTC) after receiving a failing grade. Interestingly, the original one I received was a passing grade, but how I got a second but failing grade later on in the final report card was baffling. I never told my parents about this and did not try to question the higher-ups. I found out later that the report card of another student (a habitual absentee who received a failing grade) with the same last name as mine and a similar first name initial was intentionally switched to make it appear that he passed the course. That person, I should add, was politically influential and knew a lot of officials in the higher echelon of the school. In addition, it is worth emphasizing that most high-ranking officers in ROTC were mestizos, genetic products of the Spanish and American colonization of the Philippines.

During my years in medical school, I was an unknown and average student. I was not a loudmouth and pompous person, and did not possess any political clout whatsoever. Outside of biochemistry, physiology, anatomy, pathology, and physical diagnosis, the other subjects did not stimulate my cerebral cortex. However, during the latter years and during the preparation for the national medical board examination, I was forced to pay attention to other subjects (which I carelessly ignored), and much to my surprise, I realized that I missed a lot in terms of making myself a competent future practitioner of the profession. Therefore, I used a "full-court press and pressure defense" (as in basketball) to prove myself. Within a few days after the grueling examination, my oldest sister and a very close friend and classmate of mine were both notified one early morning that I was a topnotcher and was ranked third, behind two classmates who were awarded decorative Latin words, "magna cum laude," attached to their names after graduation. A few days after the announcement was made, my rank was dropped precipitously to 9. Unfortunately, I was foolishly naïve to investigate this matter and went on to study more for the Educational Council for Foreign Medical Graduates (ECFMG) examination. Passing this examination was a ticket to enter postgraduate medical education in the US. Three decades later, I e-mailed a classmate and inquired out of curiosity about what happened to the national board ranking and why I was dropped to 9 (from 3). He told me that the record showed I was number 10! He then went on and sent me belated congratulations. That classmate, I was told, was a member of the school fraternity, and the classmates that took my spots (3 and 9) in the rank were sorority members. Hallelujah, praise the Lord. Thank you, Jesus.

Back in those days, I was aware of the existence of the strong political clout that fraternities and sororities could exert on the higher ups. Although I did not have concrete evidence to prove my point and allegations, my sister swore without a smidgen of doubt, after that phone call, that I was ranked number 3. To cut the story short, it was rumored that the test questions were manipulated and rigged to put me down (according to reliable sources). And now you know the rest of the story (a quote from the late Paul Harvey, an American radio broadcaster for ABC). It was obvious that the people in the higher echelon of the medical school, acting in concert with members of the student organization, could not accept the unexpected outcome of the board examination that an "unknown and unheralded" person like me outshined their protégés.

Why such practices would occur in clergy-run institutions are examples of contradictions and hypocrisies in a religion whose self-serving members go to church regularly and confess to wash their soul every year.

Lord, please forgive us our trespasses, as we forgive those who trespass against us, and please deliver us from evil. Amen. ❧

CHAPTER 7

Waking up to reality

"Science is the poetry of reality"
– *Richard Dawkins*

"Religion can never reform mankind because religion is slavery. It is far better to be free, to leave the forts and barricades of fear, to stand erect and face the future with a smile. It is far better to give yourself sometimes to negligence, to drift with wave and tide, with blind force of the world, to think and dream, to forget the chains and limitations of the breathing life, to forget purpose and object, to lounge in the picture gallery of the brain, to feel once more the clasps and kisses of the past, to bring life's morning back, to see again the forms and faces of the dead, to paint fair pictures for the coming years, to forget all Gods, their promises and threats, to feel within your veins life's joyous stream and hear the martial music, the rhythmic beating of your fearless heart. And then to rouse yourself to do all useful things, to reach with thought and deed the ideal in your brain, to give your fancies wing, that they, like chemist bees, may find art's nectar in the weeds of common things, to look with trained and steady eyes for facts, to find the subtle threads that join the distant with the now, to increase knowledge, to take burdens from the weak, to develop the brain, to defend the right, to make a palace for the soul. This is real religion. This is real worship".
– *Robert G. Ingersoll*

> "Information defines your personality, your memories, your skills."
> – *Dr. Ray Kurzweil, American Scientist, inventor, and futurist.*
>
> "Atheism by itself is, of course, not a moral position or political one of any kind; it is simply the refusal to believe in a supernatural dimension"
> – *Christopher Hitchens*

I emigrated from the Philippines to the US in 1971 after passing the ECFMG. Coming from a predominantly Catholic country, I had the mindset that Catholicism is "the religion" from which God's teachings originated and must be followed or else you go to hell. As soon as I started working in a hospital in Akron, Ohio, I found myself working alongside fellow physicians and members of the paramedical team of various ethnicities who belonged to different religious groups and denominations. Covert and sickening negative comments and criticisms, bigotry, and backstabbing among each other occurred sporadically early on but were soon replaced by respect, understanding, and camaraderie as the process of assimilation with American culture progressed harmoniously, despite the differences in philosophy and ideology.

Living in this great multicultural country is analogous to occupying a room with several windows, each with a panoramic view of the environment surrounding the house. A person becomes broad- and open-minded and receptive to other ideas. This is in contrast to a room with only one window, whose occupant has a constricted view of the world. Unfortunately, some had chosen, for no fault of their own, to live in a room with a limited view of the world and remained adherent to religious beliefs that were inculcated into their susceptible and gullible mind by their parents at the time when their central nervous system was young and vulnerable to indoctrination.

When I was doing my residency in the specialty of neurology in a large medical center run by Catholic nuns in the early 70s, I came across a New York City newspaper article that discussed the possibility that Jesus Christ was a supernatural alien who possessed a powerful healing power and performed miracles and "walked" on water. Such claims had roots in a worldwide bestseller book titled "Chariots of the Gods?" written by Erich von Daniken, who subsequently wrote several more very intriguing books whose themes were all about extraterrestrials who visited the earth thousands of years ago and influenced the development of early inhabitants of the planet earth. Very obviously, von Daniken and all other supporters of the theory were hit with a barrage of sharp and vile criticisms from religious zealots and pundits. Subsequently, several more books were published, all supportive of von Daniken and his associate, Giorgio A. Tsoukalos, publisher of Legendary Times Magazine. Despite all the uncomplimentary criticisms coming from zealots, it is a fact that there are numerous unsolved mysteries that occurred in the distant past on our planet. Unfortunately, those who are satisfied and fixated with the status quo (laypersons and some scientists and nonscientists) have turned a blind eye to the reality that we live in this vast universe with billions of stars, billions of light years away from each other, some of which have been in existence before the earth and our sun were born.

Did the gods from outer space, in their "chariots," travel all over the world and were venerated and worshipped as supreme gods from heaven? In the book "Ye Gods," Anne S. Baumgartner compiled a dictionary of gods described all over the ancient world. They were venerated and worshipped. Some were feared. Some were believed to be the "creator" of all humans and the earth. Aten and Marduk were the sun gods of the Egyptians and Babylon, respectively. In Turkey, Jar-Sub is considered the God of the Universe. The Mongols describe a God of the Sky, known as Tengri. In India, there is a supreme creator of the universe, Mahaskti. Zeus is the Greek god of the sky and the guardian of law and the upholder of morality.

Many would dismiss such "gods" as folklore, legends, myths, or products of imagination by the ancients. But why were they seen and worshipped all over the world? Here's the reality that everybody should agree on. Nobody knows how the universe, along with myriads of inhabitants on earth, came into existence. The claim that heaven and earth were created in just a few days, allegorical or not, is naïve and ridiculously absurd. The laws of physics tell us that the universe is very likely teeming with life. Believers would argue that in this vast universe, we are alone. Such a claim is, of course, egotistical and narrow-minded. ∞

CHAPTER 8

The skeptics, the nonbelievers, religious illiteracy, and famous quotes from famous people about religion

When one questions the validity of "accepted truths" in religion, he or she is likely to be attacked physically and mentally using deadly weapons, venomous criticism, and isolation. Despite the believers' claims that their soul is clean by being religious, and with assurance that they have a guaranteed place in heaven, the other aspect of religion tells a different story. Wars were and are being waged continuously, as we speak, all over the world because of religion. Wars between Catholics and Protestants and between Christians and Muslims, together with ethnic cleansings committed by totalitarian and dictatorial forms of governments, have killed thousands and thousands of human beings. Hitler, a Catholic, drenched his hands with the blood of millions of innocent Jews he killed during World War II.

It is true that religion leads a path to unity, but it can also lead to bigotry, hatred, and discrimination. Moreover, the unity, mind control, and mandates from religious establishments can be counterproductive and can generate controversies, defiance, and disobedience. The mile-wide gap in social and economic status between the rich and the poor has always been the source of never-ending envy, hatred, and animosity, particularly in poor catholic nations. Deaths from famine, earthquakes, tornadoes, hurricanes, microbial and viral pandemics, and other natural calamities have not been prevented by prayers.

The hypocrisies that beset most religions have spawned a number of skeptics and nonbelievers to speak their minds, further increasing the immiscibility of science and religion.

Here are some quotes from famous freethinkers, various intellectuals, scientists, inventors, politicians, writers, statesmen, celebrities, philosophers, and past American presidents about religion and God.

• **Michael Shermer** – American science writer, historian of science and executive director of the Skeptic Society – *"I'm a skeptic not because I don't want to believe but because I want to know."*

• **Dave Barry** – American author, columnist, and nationally syndicated humor columnist for Miami Herald – *"The problem with writing about religion is that you run the risk of offending sincerely religious people, and then they come after you with machetes."*

"People who want to share their religious views with you almost never want you to share yours with them."

"If there really is a God who created the entire universe with all of its glories, and He decides to deliver a message to humanity, He will not use, as His messenger, a person on cable TV with bad hairstyle."

• **George Carlin** (1937–2008) – American standup comedian, social critic, and actor – *"Religion convinced the world that there's an invisible man in the sky who watches everything you do. And there are 10 things he doesn't want you to do or else you'll go to a burning place with a lake of fire until the end of eternity. But he loves you! He loves you, and He needs money. He always needs money. He's all-powerful, all-perfect, all-knowing, and all-wise, somehow just can't handle money!"*

"I don't know how you feel, but I'm pretty sick of church people. You know what they ought to do with churches? Tax them. If holy people are so interested in politics, government, and public policy, let them pay the price of admission like everybody else. The Catholic church alone could wipe out the national debt if all you did was tax their real estate."

"Something is wrong, war, disease, death, destruction, filth, poverty, crime, torture and the ice capades. If this the best God can do, I am not impressed. This is not what you expect to find on the resume of a supreme being. It's what you expect from an office temp with a bad attitude."

"How come when it's us is an abortion and when it's a chicken is an omelet."

• **Richard Dawkins** – Author of God Delusion – *"Science has eradicated small pox, can immunize against most previously deadly viruses, can kill most previously deadly bacteria. Theology has done nothing but talk of pestilence as the wages of sin"*

"Faith can be very very dangerous, and deliberately to implant it into the vulnerable mind of an innocent child is a grievous wrong"

"I am often accused of expressing contempt and despising religious people. I don't despise religious people, I despise what they stand for"

• **Bill Gates** – American billionaire – *"Just in terms of allocation of time resources, religion is not very efficient. There's a lot more I could be doing on a Sunday morning."*

• **Sam Harris** – neuroscientist, philosopher, and author of "The End of Faith" and "Free Will" – *"Norway, Iceland, Australia, Canada, Sweden, Switzerland, Belgium, Japan, the Netherlands. Denmark, and the United Kingdom are among the least religious societies on earth. According to the United Nations' Human Development Report (2005), they are also the healthiest, as indicated by life expectancy, adult literacy, per capita income, educational attainment, gender equality, homicide rate, and infant mortality…Conversely, the fifty nations now ranked lowest in terms of the United Nations' human development index are unwaveringly religious"*

"The problem with religion, because it's been sheltered from criticism, is that it allows people to believe en masse what only idiots or lunatics could believe in isolation"

• **Robert A. Heinlein** (1907–1988) – novelist, screenwriter, aeronautical engineer, and naval officer – *"Secrecy is the keystone to all tyranny. Not force, but secrecy and censorship. When any government or church for that matter undertakes to say to its subjects, this you may not read, this you must know, the end result is tyranny and oppression, no matter how holy the motives. Mighty little force is needed to control a man who has been hoodwinked in this fashion; contrariwise, no amount of force can control a free man, whose mind is free. No, not the rack nor the atomic bomb, not anything. You can't conquer a free man; the most you can do is kill him."*

• **Carl Jung** (1875 – 1961) – Swiss psychiatrist and psychotherapist – *"The principal product of religion is guilt and despair. Despair over this life, despair over our thoughts, despair over our desires, despair over our actions. As we dwell on this despair we drive ourselves into repression in an effort to control our thoughts and desires. This repression then leads to depression, which fosters disease"*

"Religion is a cult. Everyone knows cults are bad, right? But did you know that religion shares every characteristic of a cult: authoritarian leadership, indoctrination, exclusivism, isolation, and opposition to independent thinking"

"Cults recruit people by offering a counterfeit form of acceptance and purpose, which is also how religion operates. Like a cult, religion discourages questions and marginalizes those who fail to toe the line"

Garrison Keillor – American author, story teller, voice actor, and radio personality – *"Anyone who thinks sitting in church can make you a Christian must also think that sitting in a garage can make you a car."*

• **Bill Maher** – standup comedian and television host – *"The pope is basically a cult leader."*

"I don't ridicule religion, it ridicules itself"

"Let's face it; God has a big ego problem. Why do we always have to worship him?"

• **Emma Thompson** – British actress and writer – *"I am an atheist. I regard religion with fear and suspicion. It's not enough that I don't believe in God. I actually regard the system as distressing: I am offended by some of the things said in the Bible and the Qur'an, and I refute them"*

• **Jodie Foster** – American actress – *"I cannot believe in God when there is no scientific evidence for the existence of a supreme being and creator."*

• Gloria Steinem – American feminist and journalist – *"Religion is an incredible con job."*

• **Jesse Ventura** – American politician and professional wrestler – *"Organized religion is a sham (or mind control) and crutch for weak-minded people who need strength in numbers. It tells people to go out and stick their noses in other people's business. I live by the golden rule: Treat others as you'd want them to treat you. The religious right wants you to tell people how to live."*

• **Robert G. Ingersoll** (1833–1899) – lawyer, writer, and orator – *"Christian chronology gives the age of the first man, and then gives the line from father to son down to the flood, and from the flood down to the coming of Christ, showing that men have been upon the earth only about six thousand years. This chronology is infinitely absurd, and I do not believe that there is an intelligent, well-educated Christian in the world, having examined the subject, who will say the Christian chronology is correct."*

"The doctrine of eternal punishment is the infamy of the infamies. As I have often said, the man who believes in eternal torment, in the justice of endless pain, is suffering from at least two diseases— petrifaction of the heart and putrefaction of the brain."

"A believer is a bird in a cage, a free-thinker is an eagle parting the clouds with tireless wing."

"Thousands of 'saints' have been the most malicious of the human race If the history of the world proves anything, it proves that the

Catholic church was for many centuries the most merciless institution that ever existed among men."

"I have no Protestant prejudices against Catholicism, and have no Catholic prejudices against Protestantism. I regard all religions either without prejudice or with the same prejudice. They were all, according to my belief, devised by men, all have for a foundation ignorance of this world and fear of the next. All the Gods have been made by men. They are all equally powerful and useless."

"I do not remember that one science is mentioned in the New Testament. There is not one word, so far as I remember about education—nothing about any science, nothing about art. The writers of the New Testament seem to have thought that the world was about coming to an end. This world was to be sacrificed absolutely to the next. The affairs of this life were not worth speaking of. All people were exhorted to prepare at once for the other life."

"The promise of Christ to reward those who will believe is a bribe. It is an attempt to make a promise take the place of evidence. He who says that he believes, and does this for the sake of the reward, corrupts his soul."

"Theology is not what we know about God, but what do not know about Nature."

"If a man would follow, today, the teachings of the Old Testament, he would be a criminal. If he would strictly follow the teachings of the New, he would be insane."

• **Thomas Paine** (1737–1809) – American political activist and philosopher – "To argue with a person who has renounced the use of reason is like administering medicine to the dead."

"My country is the world, and my religion is to do good."

"Of all tyrannies that affect mankind, tyranny in religion is the worst."

"The Bible is such a book of lies and contradictions there is no knowing which part to believe or whether any."

"Belief in a cruel God makes a cruel man."

"It would be more consistent that we call the Bible the work of a demon than the word of God. It is a history of wickedness that has served to corrupt and brutalize mankind."

"Priests and conjurors are of the same trade."

"The Christian religion is a parody on the worship of the sun, in which they put a man called Christ in the place of the sun, and pay him the adoration originally payed to the sun."

"All the tales of miracles, with which the Old and New Testament are filled, are fit only for impostors."

"Each of those churches show certain books, which they call revelation, or the word of God. The Jews say their word of God was given by God to Moses, face –to-face; the Christians say their word of God came by divine inspiration; and the Turks say that their Word of God (the Koran) was brought by an angel from heaven. Each of those churches accuses the other of unbelief; and for my own part, I disbelieve them all."

• **Thomas A. Edison** (1847–1931) – American inventor and businessman – *"To those searching for truth--not the truth of dogma and darkness but the truth brought by reason, search, examination, and inquiry, discipline is required. For faith, as well intentioned as it may be, must be built on facts, not fiction--faith in fiction is a damnable false hope."*

"I do not believe any type of religion should ever be introduced into the public schools of the United States."

"So far as religion of the day is concerned, it is damned fake. Religion is all bunk"

• **Abraham Lincoln** (1809–1865) – 16th President of the US, American lawyer, and statesman – *"When I do good, I feel good. When I do bad, I feel bad. That's my religion."*

• **Mahatma Gandhi** (1869–1948) – Pre-eminent leader of nationalism during the British Raj – *"God has no religion."*

• **Samuel Langhorne Clemens** (pen name Mark Twain, 1835–1910) – American writer, humorist, entrepreneur, publisher, and lecturer – *"A man is accepted into a church for what he believes, and he is turned out for what he knows."*

"The Bible is full of interest. It has noble poetry in it; and some clever fables; and some blood-drenched history; and some good morals; and a wealth of obscenity; and upwards of a thousand lies."

• **John Adams** (1735–1826) – 2nd President of the US, statesman, attorney, diplomat, writer, and founding father – *"As I understand the Christian religion, it was, and is, a revelation. But how has it happened that millions of fables, tales, and legends have been blended with both Jewish and Christian revelation that have made them the most bloody that ever existed."*

"This would the best of all worlds if there were no religion in it."

• **Thomas Jefferson** (1743–1826) – 3rd President of the US, statesman, lawyer diplomat, architect, philosopher, and founding father – *"Christianity is the most perverted system that shone on man."*

"Religions are all alike—founded upon fables and mythologies."

"The Christian god can be easily pictured as virtually the same as many gods of past civilization. The Christian god is a three headed monster: cruel, vengeful and capricious"

• **Karl Marx** (1818–1883) – German philosopher, economist, revolutionary sociologist, and journalist – *"Religion is the opium of the masses."*

• **Albert Einstein** (1879–1955) – German-born theoretical physicist and developer of the theory of relativity – *"Scientific research is based on the idea that everything that takes place is determined by laws of Nature, and therefore this holds for the action of people. For this reason, a research scientist will hardly be inclined to believe that* *events could be influenced by a prayer, i.e, by a wish addressed to a Supernatural Being"*

• **Carl Sagan** (1934–1996) – astronomer, astrophysicist, astrobiologist, cosmologist, and planetary scientist – *"You can't convince a believer of anything. For them belief is not based on evidence. It's based on a deep-seated need to believe."*

"I don't want to believe. I want to know."

"If you want to save your child from polio, you can pray or you can inoculate...try science."

• **Stephen Hawking** (1942-2018) – British theoretical physicist and cosmologist – *"Though we feel we can choose what we do, our understanding of the molecular basis of biology shows that biological processes are governed by laws of physics and chemistry and therefore are as be determined as the orbits of the planets."*

"One does not have to appeal to God to set the initial conditions for the creation of the universe, but if one does He would have to act through the laws of physics."

"Science is increasingly answering questions that used to be the province of religion."

• **Isaac Asimov** (1920–1992) – American writer and professor of biochemistry – *"Properly read, the Bible is the most potent force for atheism ever conceived."*

"The Bible must be seen in a cultural context. It didn't just happen. These stories are retreads from stories before. But, tell a Christian that—No, No! What makes it doubly sad is that they hardly know the book, much less its origin."

• **Carlespie Mary Alice McKinney** – essayist and business professor – *"Religion does three things: Divides people, controls people, deludes people."*

• **Susan B. Anthony** (1820–1906) – American social reformer and women's right activist – *"I distrust those people who know so well* *what God wants them to do because it always coincides with their own desires."*

• **Arthur C. Clarke** (1717–2008) – British science fiction writer, futurist, inventor, undersea explorer – *"Religion is the most malevolent of all mind viruses."*

• **Benjamin Franklin** (1706–1790) – American author, political theorist, politician, scientist, inventor, diplomat, and founding father – *"Lighthouses are more helpful than churches."*

• **Ernest Hemingway** (1899–1961) – American novelist, short story writer, and journalist – *"All thinking men are atheists."*

• Napoleon Bonafarte (1769 – 1821) – French military and political leader – *"All religions have been created by men."*

• **Victor Stenger** – author of "God: The Failed Hypothesis" – *"Science flies you to the moon. Religion flies you into buildings."*

It's been said that the Bible is the word of God, but to the freethinkers, the Bible is the word of several men. It's been edited, re-edited, structured, and re-structured. "I wonder what God would be like if He really existed. A man said, let there be gods and there were gods and we made them in our image."—Pasqual S. Schiavella, author of Hey! IS that you, God? To the believers, the statement is heresy and preposterous, but to the freethinkers and to those who do not want to be deluded by blind faith, it is reality and an eye-opener. The accuracy of the Bible has not been challenged until the publication of The Bible Fraud, written by Tony Bushby, an Australian writer. In this book, there were a number of provocative and challenging expositions discussed about the true identity of a man, which believers called Jesus Christ, the son of God the Father, who also has another form called the Holy Spirit. In 1415, the church of Rome ordered the destruction of two 2nd century Jewish books, Mar Yesu and the Book of Elxai, that revealed the true identity of Jesus Christ. Later, the Talmud, the authoritative source of the central text of Rabbinic Judaism and the main source of Jewish religious law, and several copies of the old testament were also destroyed. Fortunately, many copies survived, which have provided opposing information about the person called Jesus Christ. Likewise, some British documents, which recorded the true name of Jesus Christ, also survived and are kept in the British museum. These treasured documents are called the "chronicles and the Myvyean Manuscript." Richard Dawkins, author of the book "The God Delusions" wrote, "Although Jesus probably existed, reputable biblical scholars do not in general regard the New Testament (and obviously not the Old Testament) as a reliable record of what actually happened in history, and I shall not consider the Bible further as evidence for any kind of deity."

Here's some interesting historical information presented by Tony Bushby. "The Gospels of Matthew and Luke stated that Jesus Christ was the first born of Mary and Joseph and he had four younger brothers and at least two sisters (Mark 6:3). Roman Catholics are obliged to hold the opinion that the brothers and sisters of Jesus Christ were the children of Joseph by a former marriage. However, it was clearly recorded that Joseph had sex with Mary after the birth of Jesus. The statement in the Gospel of Matthew that Joseph "knew her not until she had born a son" (Matt. 1:25) eliminated the church's claim that Mary was a perpetual virgin. From the statements in the Gospels of Mark and Matthew it was clear that the brothers and sisters of Jesus were subsequent children of Mary in the fullest sense." According to Jewish records, the presentation of the virgin birth of Jesus Christ may be different from what the Catholics claim. There was a widespread belief, according to the Talmud, that Christ was an illegitimate union between his mother and a Roman soldier named Tiberius Julius Abdes Pantera. In many Jewish references, Jesus was referred to as "ben Pantera", "ben" meaning "son of." Bushby stated that the above information was obtained from the official Christian and biblical texts, Celtic annals, British chronicles, and chivalric archives of Europe. During the construction of a railroad in Bingerbruck, Germany in October 1859, tombstones for nine Roman soldiers were accidentally discovered, and one of the tombstones was that of Tiberius Julius Abdes Pantera. The inscription reads, "Tiberius Julius Pantera from Sidon, aged 62 years served 40 years, former standard bearer (?) of the first cohort of archers lies here." The tombstone is presently kept in the Romerhalle museum in Bad Kreuznach, Germany. Some believers and biblical scholars, however, think that the story of Pantera is fanciful with very little evidences and historical facts. Nevertheless, given all these historical uncertainties and controversies, it is appalling that a religion based on Jesus Christ who is considered the son of God—an almighty and supernatural being—was born I am unsure how many Christians are aware of such disclosures albeit controversial. I brought this up to my e-mail group three years ago, but I received no response whatsoever, and neither from two classmates who attended the seminary prior to entering the medical school. During my years in parochial schools, this information was not taught to the students.

In the past, there were proposals to improve religious literacy by teaching comparative religions to youths of America (William Lewis Edelen, from The Book Your Church Doesn't Want You to Read by Tim C. Leedom). Unfortunately, these proposals were met with skepticism and strong resistance by religious establishments who asserted that "all that is needed is Jesus Christ for he is the Way, the Truth and the Light. That's all our young people need." Edelen added that these same words and mythological formulas were used by Zoroaster, Buddha, and Lao Tzu of Taoism. They were essentially borrowed from Mithraism, Zoroastrianism, Egypt, Babylon, and the Greek mystery religion."

If you are bad, you go to hell, but where the hell is hell? Albert Einstein (1879–1955) remarked, "I cannot imagine a God who rewards and punishes the objects of his creation, whose purposes are modeled after our own—a God, in short, who is but a reflection of human frailty."

Punishing the object of His creation is analogous to destroying your computer when it won't do what you want it to do after programming it. Upon death, the faithful believe that there is a soul that separates from the body. What is the composition of the soul? They believe that the soul represents the mind and emotion, but it is without organic or physical composition. If such is the case, how can it burn in the eternal flame of hell? The belief in the existence of a soul that separates from the organic body at the time of death has its roots in near-death and out-of-body experiences, which, in my opinion, are generated by abnormal cerebral activities that take place in the parietal-temporal region when the brain's metabolic system is compromised by a lack of blood flow, infection and inflammation, hemorrhage, or epileptic spells. This phenomenon can be reproduced by stimulating the parietal-temporal area of the brain. The intrusion of rapid eye movement (REM) sleep into the subconscious, a stage in sleep that follows an initial phase of sleep (or non-REM sleep), causes the individual to develop intense dreams wherein the individual is unable to move (sleep paralysis). These dreams may consist of a sensation of falling, floating, and dissociation of the consciousness and the physical body.

Atheists and agnostics are vilified, chastised, shunned, ostracized, and are even considered followers of the devils by the believers. One would wonder the numerous wars waged by religious people against each other, not to mention the hypocrisies and contradictions that bedevil and plague their organizations since time immemorial. I recall during my childhood when my relatives would mock and make fun of members of other religions while attending their religious processions. Children were also ordered to turn the radio off or to change the station immediately (there was no television during that time) when non-Catholic gospel songs were played. The vengeful, jealous, and cruel God of the Old Testament that annihilated pagan nations was not part of religious education, and the only emphasis was on the kinder, gentler, and loving God of the New Testament— the same God that will throw "sinners" down to the eternal flame of hell.

Some Christians may not know the origin of the word "pagan." It all began when Emperor Constantine persecuted those who refused to follow Roman teachings. He offered bribes to those who joined the new religion of Jesus Christ and granted freedom to slaves. He issued edicts prohibiting followers of other belief systems from congregating. Their scholars and philosophers were silenced, and their works burned. Those who refused to accept Roman rule were forced to flee to remote regions of Pagi to avoid persecution. They subsequently became known as pagans. As part of the campaign of the Roman Empire, they were vilified, dehumanized, treated as people without religion, and labeled as heathens. Realistically, the Romans and the Pagans worshipped the same deities, but many Christians denounced Paganism as a false religion. If such was the case, Christianity is also false.

"I like your Christ, I do not like your Christians. Your Christians are so unlike Christ." – *Mahatma Gandhi*

"I say deliberately that the Christian religion, as organized in its churches, has been and still is, the principal enemy of moral principles in the world." – *Bertrand Russell* ❧

CHAPTER 9

UFOs, flying saucers, ancient spacecraft and Vimanas, the space vehicle of Ezekiel, and da Vinci's paintings

"The key to the future lies hidden in the past"
– *Erich von Daniken*

"Knowledge of the past and of the places of the earth is the ornament and food of the mind of man"
– *Leonardo Da Vinci (1452 – 1519), Italian polymath, painter, engineer, and scientist*

"To my way of thinking, there is every bit as much evidence for the existence of UFOs as there is for the existence of God, probably far more. At least in the case of UFOs, there have been countless tapes and films and, by the way, unexplained sightings from all over the world, along with documented radar evidence by experienced military and civilian operators"
– *George Carlin*

Believers have a unique stereotypical mental paradigm that in this vast immeasurable universe consisting of countless galaxies and billions of stars and planets, humans are the only inhabitants, and outside of Earth, there exists no one else. Thus, the sightings of unidentified flying objects (UFOs), putatively of extraterrestrial origin, are considered hogwash. Such is their fixed mindset. There is, however, pictorial and descriptive evidence of UFOs and space vehicles that have activated the optic and occipital neurons of both ancient and modern humans. The first report of a sighting was recorded on June 24, 1947, when a pilot, Kenneth Arnold, spotted nine disc-shaped objects flying at least 1220 miles per hour. At that time, similar sightings also occurred in Oregon and Idaho and several hundred more in July. He described the objects as similar to pie plates, hence the name "flying saucers." In 1974, a book titled "The Spaceships of Ezekiel" was published by Josef F. Blumrich (1913–2002), an Austrian-born aeronautical and aerospace engineer who was involved in NASA's Systems Layout Branch Program at the Marshall Space Flight Center from 1959 to 1974. Modern technological procedures were used to reconstruct the strange flying vehicles that descended from the sky and were witnessed by one of the four Jewish prophets, Ezekiel, several times for several years, 2500 years ago. In von Daniken's "Chariots of the Gods?", encounters with sky vehicles were interpreted as alien visitations. In an effort to prove that von Daniken was wrong, Blumrich was prompted to carefully read the biblical Book of Ezekiel. The description was as follows: "Their legs were straight, and the soles of their feet were round, and they sparkled like burnished bronze." Ezekiel's account of the structures of the wheels of the vehicle was as follows: "They sparkled like chrysolite, and all four wheels looked alike. Each appeared to be made like a wheel intersecting a wheel. As for the likeness of the living creatures, their appearance was like burning coals of fire and like the appearance of lamps: it went up and down among the living creatures; and the fire was bright, and out of the fire went forth lightning. I looked, and I saw a windstorm coming out of the north—an immense cloud with flashing lightning and surrounded by brilliant light. The center of the fire looked like glowing metal." Blumrich later agreed with von Daniken that the sky vehicles Ezekiel saw were indeed extraterrestrial.

Throughout history, reports of powerful people and members of royalties travelling and hopping from different cities have included

Solomon and Arabians on flying carpets. Flying chariots in China and flying aircraft in ancient India were recorded in their ancient literature. The flights were considered myths and legends. In India, the aircraft were called "Vimanas," as depicted in two famous ancient texts, Ramayana and Mahabharata. According to the Indian scholar Ramachandra Dikshitar (Vimana Aircraft of Ancient India & Atlantis by David Hatcher Childress), there were other Indian texts that mention these flying machines UFO encounters in the Far East have been reported and recorded in ancient Chinese writings (The Chinese Roswell, Hartwig Hausdorf, 1998).

"If flying saucers exist, if they are in fact space vehicles from another world, and if they have the performance capabilities described in various UFO reports, then there is no doubt that these vehicles would have provided adequate transportation for any beings who might have been involved in nourishing the Biblical religion" (The Bible and Flying Saucers by Barry Downing). Speculations emerged that Israelites were guided by UFOs during their travel from Egypt to the Red Sea. The UFOs appeared cloud-like during daytime and glowed in the dark. The parting of the Red Sea, which to date remains a mystery, might have been the work of extraterrestrials inside their space vehicles, according to alternative theorists.

The paintings of Leonardo da Vinci, the genius and a polymath of medieval times, are quite intriguing, to say the least. He once remarked, "learning never exhausts the mind." The portrait of Mona Lisa might have a hidden face of an alien when mirrored. The portrait of St. John baptizing Jesus Christ shows a glowing disc object hovering in the sky. Similar flying objects were seen in the sky above the cavalry, where Jesus was crucified. Believers and skeptics obviously avoid further discussion on the subject.

Stanton T. Friedman (1934–2019) was a Canadian nuclear physicist and the original civilian investigator of the Roswell UFO incident. He was the author of "Flying Saucers and Science". To date, UFO believers are considered cranks and weirdoes, but such attitudes are slowly dissipating, thanks to the pioneering work of

Friedman. Despite the general consensus that interstellar travel would require too much energy and time, most believe that we are not alone in this universe. Support for the existence of UFOs has increased, thanks to Tucker Carlson's Fox News segment on the subject. Carlson was astounded at what was recorded and kept secret by the US military. The Navy had confirmed that the videos captured hundreds of sightings from 2004 and 2015. They were published online and shown on Fox Channel numerous times. Pressure from the public, including Nick Pope of the British Defense Ministry, has led to the creation of a bipartisan commission to investigate UFOs. Lately, however, the Pentagon UFO Commission has been criticized for lack of transparency. It is quite likely that military officials do not have enough data and information to discuss with the public.

Dr. Michio Kaku, a well-known theoretical physicist, says, "The US government's admission of a UFO means there are top-secret military weapons in use...or aliens way smarter than us are really here."

Documented scientific proof of alien visitation in Japan was offered by Vaughn M. Greene, who was in the Army of Occupation, Japan, during the Korean War and worked for the Ryan and Convair Aircraft Company and for the navy under the Department of Defense. His research activity led to the publication of the book titled "The Six Thousand Year-old Space Suit". In this book, he analyzes several statues known as "Dogu." He presented several similarities between the statues and modern space suits. In addition, Greene studied Japanese legends, Shinto mythology, and ancient folklore, all of which described beings from outer space. Dogu varied in height from 3 to 12 inches and were built 10,000 BC to over 500 years B. The Jomon people, who were a crude Neolithic people of Japan, were the first people on earth to make ceramics, the building material of Dogu.

The following are legitimate questions that deserve reasonable scientific discussion: Is it possible that religion and belief in an invisible God evolved and was groomed by intelligent and superior beings who have been on earth since the beginning of time? Where did they come from? Do human genes have extraterrestrial origins? Who were Jehovah and Yahweh? Why were there so many gods or supreme beings revered by many people in different parts of the world in the distant past? Were they extraterrestrials who travelled all over the world in their space vehicles to educate (or rule) the Earthlings? Were angels, the messengers of God, beings with jet packs that enabled them to travel from one place to another? Were Greek gods mythical, or were they extraterrestrials? According to von Daniken, they were powerful beings who travelled across the sky in their chariots. They threw "lightning bolts." Is it possible that these superior beings may still be around and are monitoring human development? Are these questions far-fetched? Maybe yes, maybe not. ❧

CHAPTER 10

The Sumerians and the account of creation. Who do you believe?

> In 2007, *Christopher Hitchens* (1949–2011), a British-American author, columnist, orator, and journalist, and the author of "God is not Great: How religion poisons everything", wrote, "God did not make us, we made God. Religion is a distortion of our origins, our nature, and the cosmos. We damage our children—and endanger our world—by indoctrinating them."
>
> "In spite of the land's natural drawbacks, they turned Sumer into a veritable Garden of Eden and developed what was probably the first high civilization in the history of man"
> – *Samuel Noah Kramer, The Sumerians: Their History, Culture, and Character*
>
> "120 million of us place the big bang 2,500 years after the Babylonians and Sumerians learned to brew beer."
> – *Sam Harris*

According to the Bible, God created the universe 6000 years ago, in six to seven days. The believers in the supernatural will not accept or entertain any other dates. However, scientific evidence collected using radioactive elements shows that the Earth and the universe are 4.5 and 13.5 billion years old, respectively. The origin of life on Earth remains unknown, but suffice it to say that there are controversies abound on how, when, and why life began on Earth. Who did God create first: a white Caucasian, a Black, or an Asian person? If he created a white person first, did he travel to Africa and Asia to create Black and Asian people, respectively? What about the Mesopotamians and Mesoamericans? What does the Bible say?

Poisoned arrows continue to fly in all directions as proponents of extraterrestrial origins of man, along with a Sumerian account of creation versus the proponents of supernatural creation, who try to stand their grounds. Zecharia Sitchin (1920–2010), an Israeli-American writer, claimed that the ancient Sumerian culture was created by Anunnaki gods ("An" or "Anu" means sky, and "Ki" means earth) that inhabited a planet beyond the orbit of Neptune, called Nibiru, with a periodic elliptical orbit around the sun every 3600 years. They came to earth 400,000 years ago, in search of gold, which they needed to sustain life in Nibiru, and established the earliest and an advanced civilization in Mesopotamia, or "lands between rivers," 5500 years ago. These rivers are Euphrates and Tigris. The region of Mesopotamia is now occupied by Iraq, Iran, Turkey, and Syria, and the land is known as "Fertile Crescents and the Cradle of Civilization."

Sitchin, based on his interpretation of Sumerian-Babylonian cuneiform tablets, along with Alan F. Alford (1961–2011), a British writer and author of the "Gods of the New Millennium", claimed that Anunakis or the sky gods created humans in their image 200,000 years ago. They also built the pyramids, the Sphinx, and other stone sites. They probably had a reptilian appearance, according to R.A. Boulay, the author of "Flying Serpents and Dragons: The Story of Mankind's Reptilian Past."

The Sumerians were credited for their inventions of time by dividing day and night into 12-hour periods, hours into 60 minutes, and minutes into 60 seconds. They established a form of government, agriculture, architecture, science, astronomy, and architecture. How they acquired this knowledge is intriguing. This account of creation became one of the reference points and inspirations for many ancient alien theorists, including Daniken, Tsoukalos, and many others.

Despite the evidence provided by Sitchin, his detractors—Michael Heiser, a well-known biblical scholar and author of several books who adamantly refutes the existence of extraterrestrials, and Jonathan Gray, an archeologist and author of "Dead Men's Secret"—do not speak highly of him. They think Sitchin misinterpreted the Sumerian chronology, and they think, "Sitchin simply got it wrong." No further debates ensued between them as Sitchin died at the age of 90. As expected, one of his staunch supporters, Aerik Vonderburg, author of "The Genesis of Revelation", maintains his support for the extraterrestrial deity hypothesis. The groundbreaking findings he discussed in his book exposed the misunderstandings and misinterpretations of the Bible about Jesus Christ and God. According to Vonderburg, Heiser's strong antagonism toward Sitchin is based on his Judeo-Christian faith together with his "immature, mean-spirited, and arrogant behavior." Moreover, he totally discounted the descriptions of flying objects or Vimana aircrafts in ancient India and the recent sightings of UFOs by US and Russian militaries. Nick Pope, a British journalist who has been a frequent Fox News contributor, is a leading modern UFO researcher. Michio Kaku, a well-known theoretical physicist, indicated that the beings visiting earth in their UFOs are clearly and highly technologically advanced.

Two years ago, I questioned my overly religious classmates about why God chose the Mideast to be the birthplace of Jesus Christ and the Virgin Mary to be his mother. Why not a Filipina, a Chinese, or a Japanese? Why the choice of a white Caucasian? Why not the aborigines of the Philippines or Australia, or a black African woman? Sitchin and Alford might be right. Modern humans were crossbred with primitive earthlings at least 200,000 years ago via genetic manipulation in the so-called "Garden of Eden." In the ancient Sumerian illustrations, serpentine helical structures, which can be interpreted as the DNA helix, were shown. What does the staff of Asclepius, a physician symbol, remind you of?

In the past five decades, tremendous advances in DNA technology have changed the way modern humans view the account and mechanism of creation. In vitro fertilization is now commonplace, and is indicated in situations when a woman has local pathology in the fallopian tubes. In this case, a female cell is fertilized by a male cell in a culture medium to form a zygote, and later implanted into the woman's womb. Reproduction without sexual intercourse is now possible using the process of cloning. The process involves extracting the DNA, using a needle, from a mature somatic cell (a skin cell or any adult cell) and injecting it into an egg cell that was previously emptied of its own DNA. The "new" cell is then allowed to develop in the culture medium and later implanted into the womb of another animal. At this time, however, cloned embryos have short life span and, in general, their organs do not develop maturely in contrast to the organs of traditional embryos. The first animal cloned was a sheep named Dolly. It survived for six years (in contrast to the sheep's average life span of 12 years). Reports of human cloning have not been substantiated despite claims from researchers in Asia. Cloning and gene therapy, when perfected, will soon provide cure for those incurable genetic and degenerative disorders including cancer. It should be emphasized that these biological technological advances did not take place with the help of prayers or the Bible.

Regardless of who created humans, so many questions remain to be answered. Why is there variability in skin color and races among humans? Although this variation can be explained by molecular genetics and mutations, how it came about is another issue. The Bible will obviously provide no answer. For believers, however, "the buck stops at the Bible and God," no questions asked and just believe what you were told. ଓ

CHAPTER 11

Unexplained and mysterious events and creatures misinterpreted as divine, miraculous, and supernatural.

"The study of theology, as it stands in Christian churches, is the study of nothing; it is founded on nothing; it rests on no principles; it proceeds by no authorities; it has no data; it can demonstrate nothing; and it admits of no conclusion."
– *Thomas Paine*

"Just because you believe in something does not mean it is true."
– *Albert Einstein*

"They told me to use the brain God gave me. I did. Now I'm an atheist. Ironic, isn't it?"
– *Atheist Saying, www.geckoandfly.com*

According to the Cambridge Dictionary, apparition is the spirit of a dead person appearing in a form that can be seen. The Merriam-Webster Dictionary defines it as an unusual or unexpected sight of a ghostly figure. Among Catholics, apparitions of the Virgin Mary (not Joseph, not the saints, and not the apostles) are sporadically reported. Very few had been "authenticated" by the church, but by and large, believers consider them as revelations or a personal message and are usually associated with turmoil, catastrophic events, and wars. The incidence of apparitions in predominantly non-Christian nations is unknown, but it seems common in Catholic nations, such as the Philippines. I recall that my late uncle would travel several hundred miles to witness the reported apparitions of the Virgin Mary. He prayed, asked for forgiveness, and expressed remorse for his sins. He was a womanizer and had several children born out of wedlock. He worked for the local police department and was involved in illegal and nefarious schemes. He regularly collected money from Chinese businessmen in exchange for protection from hoodlums and street gangs, which were actually protected by the police. I was young and ignorant at the time, but I was befuddled and unable to understand such unethical and immoral acts. As the years went by, I realized that these practices were rampant in urban cities like Manila. Praise the Lord!

On a personal level, I find it difficult to believe in the existence of apparitions of the Virgin Mary, and it is extremely difficult to provide clear and direct evidence for or against their supernatural origin. Skeptics believe they are imaginations (or fabrications) of very religious people. Given the inconsistencies in the Bible, one wonders whether these biblical personalities, who were ecclesiastically considered members of the supernatural family, are real or not. No one is unable to get a handle on this subject matter, including the alternative possibility that apparitions may be of extraterrestrial origin. One thing is noteworthy: percipients of apparitions are deeply religious. To my knowledge, they have not been reported by nuclear physicists, astronomers, atheists, or non-religious neuroscientists.

In the book titled "Fatima Revisited: The Apparition Phenomenon in Ufology, Psychology, and Science" compiled by Dr. Fernando Fernandes, Dr. Joaquim Fernandes, and Raul Berenduel, a contributor by the name of Michael A. Persinger, PhD (1945–1918), a Canadian-American Professor of Psychology at Laurentian University, wrote, "One of the best documented cases of the 20th century involving luminous phenomena framed within the UFO typology occurred in Fatima, Portugal in October 1917. Approximately 70,000 people observed a sphere of unusual light. Many of these witnesses also saw several images that included variants of Joseph, Mary, and Jesus, sacred symbols of the Christian religion. This paper attempts to establish a causal relationship between these visions and the physiological and neuro-physiological alterations experienced by witnesses, as well as tectonic conditions within the Fatima region. The author theorizes that stimulation of the temporal lobe may be the actual cause of the apparition phenomenon." It is well known that various visual and auditory symptoms are mediated by the temporal and occipital lobes. These symptoms include visual and auditory hallucinations, palinopsia, and palinacousis, which can be induced by various conditions, such as tumor, infection, stroke, medications, seizures, and vascular lesions. The word palinopsia is derived from the Greek word "palin," which means "again," and "opsis," which means seeing. It is the perseveration or persistence of an image or seeing an image after the initial stimulus is removed. It is different from a physiologic afterimage that occurs transiently after staring at a bright object. Palinacousis (acousis means hearing) has the same mechanism. Certain sounds persist or are repeated many times and recur sporadically. This symptom has been reported in people with tumors or seizure focus in the temporal lobe. One will wonder how medically unsophisticated ancient people interpreted these symptoms.

Several books have described the Fatima phenomenon, with most of them describing shafts of brilliant red light emanating from the rim of the revolving luminosity associated with other colors, including green, violet, and blue. Given the religious context of the situation, many people became frightened and prayed for forgiveness and yelled, "I believe!" (True Story of Fatima by Joao de Marchim, 1952).

The Old and New Testaments are filled with descriptions of God and various "angels" who descended from the sky out of the clouds and interacted with humans. These "winged" beings accompanied the "Lord" most of the time. They acted as messengers in the name of their leader. When Moses went to Mount Sinai to meet God, there was "thunder and lightning," together with thick clouds wrapped around the mountain. For primitive and unsophisticated earthlings, those beings descending from the sky inside flying machines that emitted fire and smoke were superior and almighty. If modern humans aboard spaceships, land on an island inhabited by primitive people isolated for thousands of years from modern civilization, and travel from one place to another with their jet packs, how would the natives treat them?

Interactions between humans and various "gods" along with animism have been well documented since the beginning of time on planet earth. Interestingly, these powerful beings descended from the sky in their fiery space vehicles and made their descent with thunderous roars of lightning. Yahweh, the national god of ancient Israel and Judah, said, "I am from above." Jesus also said, "I am not of this earth." Ann Madden Jones, the author of "Yahweh Encounters: Bible Astronauts, Ark Radiations, and Temple Electronics", found in her studies of the Christian Bible various depictions of interactions between humans and advanced alien beings using electronics in the ark and tabernacle complex in Solomon's Temple, before it was destroyed during the reign of Nebuchadnezzar, the Babylonian king, in 586/587 BCE.

The Ark is the center point of Jewish worship. It contains the 10 commandments that Moses received from God, along with instructions on how to build and use the Ark, which turned out to be an enigmatic technological machine. According to the Bible, several thousand people died from misuse and mishandling of the Ark. According to ancient alien theorists, Daniken and Sitchin, Yahweh was an alien visitor who communicated with the ancients via the Ark tabernacle complex system. Some Biblical scholars believe that the Ark of the covenant was a giant electroradioactive generator or capacitor that contained and powered the manna machine to manufacture foods from algae, enabling Jews to survive during their 40-year trek in the wilderness to escape from Egyptian persecution. Its whereabouts are unknown since the first temple was destroyed. ❧

CHAPTER 12

Neuroanatomical and neurochemical bases for good and bad behavior

"I think religion is a neurological disorder"
– Bill Maher

"Human behavior flows from three main sources: desire, emotion, knowledge."
– Plato, Greek philosopher

"The men and women on death row have some combination of bad genes, bad parents, bad environments, and bad ideas (and the innocent, of course, have supremely bad luck). Which of these quantities, exactly, were they responsible for? No human being is responsible for his genes or his upbringing, yet we have every reason to believe that these factors determine his character. Our system of justice should reflect an understanding that any of us could have been dealt a very different hand in life. In fact, it seems immoral not to recognize just how much luck is involved in morality itself"
– Sam Harris

"Recent brain scans have shed light on how the brain simulates the future. These simulations are done mainly in the dorsolateral prefrontal cortex, the CEO of the brain, using memories of the past. On one hand, simulations of the future may produce outcomes that are desirable and pleasurable, in which case the pleasure centers of the brain light up (in the nucleus accumbens and the hypothalamus). On the other

hand , these outcomes may also have a downside to them, so the orbitofrontal cortex kicks in to warn us of possible dangers. There is a struggle then, between different parts of the brain concerning the future, which may have desirable or undesirable outcomes. Ultimately it is the dorsolateral cortex that mediates between these and makes the final decision. Some neurologists have pointed out that this struggle resembles, in a crude way the dynamics between Freud's ID, ego and superego."
– *Michio Kaku, theoretical physicist and professor of theoretical physics in the City College of New York.*

The brain, just like any other organs, can be compromised by various diseases, and the symptoms and neurologic deficits resulting from certain disease processes depend on the parts of the brain that are affected. The physiological and biochemical activities of extracranial organs, such as the liver and digestive system, blood cellular elements and bone marrow, and peripheral endocrine organs, separate us from one another. The ability to metabolize foods and drugs as well as the ability to dispose of the by-products of metabolism determine the resistance or susceptibility of human beings to develop certain diseases, while the integrity of blood cellular elements, particularly white blood cells and antibodies produced in the bone marrow, will determine the resistance or susceptibility to certain infectious diseases. The endocrine organs control our weight, ability to cope with stress, as well as blood pressure and heart rate. Some are susceptible to developing osteoporosis, while others are susceptible to acquiring skin cancer. Differences in genetic make-up may predispose humans to develop malignant tumors, degenerative disorders, and inherited incurable diseases.

The brain, an organ encased inside the cranial vault, determines our personality traits, such as behavior, emotion, and intellect. It stores memory, remote and recent. Some people react to stress in a different fashion. Some people can take jokes, and some cannot. Some have better memory and intellect than others. Some are optimistic, and some are pessimistic or cynical. Some are extroverted, and some are introverted. Some are gullible, and some are not. Some are impulsive and some are restrained. Some can be easily indoctrinated, and some tend to be skeptical – or resistant to indoctrination.

Lesions in the brain, such as infarction due to stroke or hemorrhage, infection and tumor, or a surgical lesion or removal of certain parts of the brain, can affect the behavior and mental function of an individual. Chemically, certain drugs, such as decongestants and antihistamine, can affect memory temporarily, while degeneration of neurons in the nucleus basalis, which contains acetylcholine, will lead to permanent loss of memory in Alzheimer's disease. Loss of dopamine, as in Parkinson's disease, will impair locomotion and coordination, while excessive dopamine excess related to overmedication in this disease can cause hallucinations. Hyper-dopaminergic activity is the chemical basis in people suffering from psychosis, e.g., schizophrenia. Anti-dopaminergic or antipsychotic drugs usually control hallucinations. Some men with excessive androgen activity are sexually active or aggressive. An imbalance in dopamine and serotonin metabolism in the brain can lead to depression and anxiety, which may fluctuate, as seen in patients with bipolar disorder. Drugs such as ethanol, steroids, cocaine, sedatives, opiates, psychedelics, antidepressants, serotonin reuptake inhibitors, varenicline (smoking cessation drug), and stimulants like amphetamines can induce violent behavior. It all boils down to alteration of the activities of various neurotransmitters such as gamma-aminobutyric acid, serotonin, norepinephrine, acetylcholine, glutamate, aspartate, glycine, taurine and dopamine. In essence, human behavior is mediated and controlled by neurotransmitters of the limbic system, and not by the so-called soul or spirit. Similarly, personality of animals is groomed and influenced by these neurotransmitters. Animal lovers know very well that kittens and puppies grow up with different personalities. Some are playful and social, some are shy, aloof, fearful and quiet, some are enthusiastic, and some are easily excitable and aggressive. Medical professionals know that bacteria and parasites have neurotransmitters and they produce various chemicals that are toxic to animal hosts. Some are pathogenic and some are not. Do these microbials have soul? If they do, will their soul go to hell after they are destroyed by antibiotics?

Bilateral removal of the hippocampus, including any neighboring structure, such as the amygdala, will lead to permanent memory loss, loss of recognition, abolition of fear, and violent behavior. Chronic alcohol intake can lead to memory loss and confabulation. Lesions in the limbic lobe, as seen in encephalitis, can lead to hallucinations, anxiety, depression, confusion, and seizures. A specific region in the brain when repeatedly activated by subthreshold electrical or chemical stimulation - also known as kindling - can cause epileptic seizures. Patients with frontal lobe lesions or following frontal lobectomy frequently develop social disinhibition, lack of spontaneity and initiative, impairment of cognitive function, apathy, disorientation, and agitation. Damage to the limbic and prefrontal cortex can affect cognition, memory, and emotional processes. Abnormal functioning of the orbitofrontal cortex, dorsolateral pre-frontal cortex, and anterior cingulate cortex can result in various personality disorders and aggression. Physical and sexual aggression, depression, paranoia, anxiety, and sleep disturbances may occur as early manifestations of Alzheimer's disease. Individuals with a history of domestic violence and alcohol abuse have lower metabolic activity in the hypothalamus, thalamus, and orbitofrontal cortex. A tumor that affects the occipital lobe can cause visual disturbances, visual loss, and, at times, visual perseveration (palinopsia). Hyperreligiosity has been reported in people with lesions in the temporal lobe and in people with psychoses. In patients with parietal lobe dysfunction, the inability to perceive tactile sensation and identify body parts, loss of visual, spatial, and configurational function, and denial of paralysis usually occur and, at times, are mistaken for malingering by the unwary.

An interesting and thought-provoking subject matter was discussed in Time Magazine dated December 3, 2007. The front cover reads, "What Makes Us Good/Evil. Humans are the planet's most noble creatures—and its most savage. Science is discovering why" (by Jeffrey Kluger, a senior writer for the magazine and author of several books, including Apollo 13, Apollo 8, and two novels for young adults). Kluger was spot on when he presented the various areas in the brain, particularly the prefrontal cortex, that mediate aggression and violence. Was it possible that Joseph Stalin, Augusto Pinochet, Adolf Hitler, Osama Bin Laden, and Pol Pot all harbored abnormal neurons and dysregulated neurochemical transmission in their prefrontal and cingulate cortexes, including the amygdala? If that was the case, whatever they had in their brain was the opposite of altruistic human beings, such as Mahatma Gandhi, Mother Theresa, Martin Luther King Jr., and The Dalai Lama. If those atrocious leaders harbored abnormal neurons and neural connections, are they now in hell and suffering endlessly in hell? If that was the case, why would God punish the objects of his (or her or its) creation. If those abnormal neural connections developed over time through years of bad parenting and bad social influence, who would be held responsible for their atrocities? Would their parents go to hell, too?

According to believers, everything is spiritual, and anything that goes against the law of morality or anybody that disobeys the 10 commandments goes to hell. Moreover, they believe that confession may guarantee a place in heaven. If Hitler, a Catholic, was alive today and confessed and recited 1000 "Our Fathers," 1000 "Hail Marys," 1000 "Apostle Creeds," and prayed using the holy rosary every day for 365 days, would he be forgiven after killing millions of innocent Jews? I guess it would have been quite unreasonable to forgive him, but then, from a scientific standpoint, was he harboring abnormal neurons in his limbic lobe that drove him to commit such a heinous act? ❧

CHAPTER 13

Contradictions, hypocrisies, paradoxes, and ironies in religion, and victims of Christian faith atrocities: Religion is an inherently violent institution

"As to the book called the bible, it is blasphemy to call it the Word of God. It is a book of lies and contradictions and a history of bad times and bad men."
– Thomas Paine

"In a theological seminary, if a professor finds a fact inconsistent with the creed, he must keep it secret or deny it, or lose his place. Mental veracity is a crime, cowardice and hypocrisy are virtues."
– Robert G. Ingersoll

"Anyone who attempts to construe a personal view of God which conflicts with church dogma must be burned without pity."
– Pope Innocent III, greatest single pre-Nazi mass murderer, 1209.

"It was even possible for the most venerated patriarchs of the church, like St. Augustine and St. Thomas Aquinas, to conclude that heretics should be tortured (Augustine) or killed (Aquinas)"
– Sam Harris

> "Christianity is the most ridiculous, the most absurd, and bloody religion that has ever infected the world"
> – *Voltaire*

Hypocrisy is defined as a pretense of having a virtuous character and moral beliefs, and being adherent of justice. Is hypocrisy a natural human behavior or part of self-preservation? On April 24, 2020, I e-mailed my classmates a list of contradictions, hypocrisies, paradoxes, and ironies in religion. Surprisingly, only five answered me. One of them declined to comment, and the other savored the moment of truth presented in the mail. The other three made side comments that were not directly addressed to me. They were probably intimidated or upset. One of them got "pissed off" and never communicated with me again.

It is noteworthy how Roman Catholicism focuses on the so-called Holy Trinity despite the belief that there is only one god who is omnipotent, omniscient and omnipresent. Therefore, mathematically speaking, it is 3 in 1 and 1 in 3. In the Catholic Church, hypocrisy dates back to when Jesus was alive but was eventually killed by either the Romans or Jews. If it was the former, it is ironical that Roman Catholicism was founded by a group of people who regarded Jesus as a threat to Roman rule. If it was the latter, one can understand why some Christians, who believe in the culpability of the Jews in the death of Jesus, still harbor the anti-Judaism sentiment instilled into their mind during the early days of religious indoctrination. And so, a religion was born out of these uncertainties, ironies, hypocrisies, and mathematical incongruities.

As I alluded to previously, it is an eerie and awkward feeling for someone coming out of the confession box after revealing his or her sins to a priest, only to find out later that priest was a pedophile and a sex abuser. In 2018, a priest was convicted of sexually abusing children during his 40 years of officiating in a church in the Philippines, a predominantly Catholic nation. Recently, in October 2021, BBC News reported a horrifying revelation that 216,000 children were sexually abused by clergies in France since 1950. Apparently, none of my religious classmates bothered discussing the news item, so I decided to post the news to the e-mail group after waiting for several weeks. One classmate who lives in the New England area answered back and accused me of being anti-religion.

The incidence of pedophilia among Catholic priests many centuries ago is unknown. It will not surprise anyone nowadays (except staunch defenders of the Catholic religion, exemplified by my classmates in premedical and medical school) that this condition (or problem) has been in existence for hundreds of years, maybe more. Some may wonder how many priests who were later canonized to be declared saints committed pedophilia.

Several years ago, when my family and I moved to upstate New York, we met a "nice" young priest after the conclusion of a Sunday mass. The priest approached my teenage boy and talked to him with a low voice while my wife and I and my other children were having a conversation with other parishioners, several feet away. Our boy spoke to us upon arrival at home and told us that he had been invited by that priest to attend a summer camp with him for one week. At that time, there was no media attention to sexual abuse on children. To cut the story short, we politely refused. We found out later that this priest was, among many others, being investigated for sexually molesting children. That priest had since been expelled from the diocese.

Here's an experience that my daughter wanted me to share with the readers of this book. One day, she was attending a confirmation in a local parochial school. When it came to her turn in line for confirmation, she stepped up and smiled at the bishop. He returned the smile and said, "Before we start, can I take this one with me?" She was "thankful," but that was to be her last active participation in the Catholic Church. That priest, a high-ranking member of the Catholic diocese in the Capital District of New York, was later accused of molesting an 11-year-old boy during the 70s. He was also accused of covering up for other priests in New York who sexually abused young boys for decades!

Euphemisms and political correctness are now part of our present-day society. Terms such as sexual preference or orientation and minor-attracted person are now being used to avoid bloodshed and violent protests. Some traditionalists from all sectors of society do not have respectful or complimentary attitudes toward homosexuals. However, some neuroscientists and neuropsychologists believe that there is a scientific basis for homosexuality, and even for pedophilia. At this time, however, sexual abuse of children is a crime (or a sin), but what if there is a yet-to-be-discovered neuroendocrine, neurophysiologic, and neurochemical explanation for or aberration in the brain of pedophiles?

During my early teens, I had the occasion of joining my older cousins in grooming roosters for cockfighting, a lucrative sport and a national obsession in the Philippines. It is a big-time gambling avenue and a huge source of income for some Filipinos. Two roosters, armed with pointed, razor sharp blades attached to their legs, fight to their bloody death. The dead loser rooster would then be brought home by the owner of the winner rooster (along with thousands of cash) and eaten. Who gave the very religious Filipinos the right to let fighting roosters kill each other? This is an egregious cruelty to animals committed by religious people living in a predominantly catholic nation. Where are the animal rights activists?

People pray for the safety of others from tragedies, famine, diseases, microbial pandemics, wars, and natural disasters, but people die anyway. Still, religious people believe that God loves us. Religion, the followers are told, is a good and peaceful thing, but how about wars between Catholics and Protestants in Ireland and those between Christians and Muslims? How many were waged by the agnostics or atheists in the past? We were told that God created the universe in six days, six thousand years ago. But what about the billions of galaxies, billions of light years away, that had been in existence before the earth and the sun were born. Jesus told us to be humble, honest, and to live a simple life. However, many religious people who go to church regularly cheat on their taxes, make obscene profits from their business and stocks, and live extravagantly. In the Philippines, the rich don't care about the poor living in slums and squatter areas in Manila. They drive luxury cars and make fun of people driving cheap cars. Nice, honest, and decent Filipinos call them a bunch of snobs—the same people who go to church and pray every Sunday.

During this modern time, advances in medical science occur at a fast rate. What were once unidentified and incurable diseases are now well defined and treatable. If a nice, peace-loving person receives a heart or kidney transplant or any other tissues or organs from a violent brain-dead serial killer, will the recipient, who is now able to live because of those transplanted organs, go to hell when he dies? Conversely, if that serial killer receives organs from a benevolent brain-dead, sinless person, will the former go to heaven or hell when he dies? If that sinless person dies, what will happen to his soul and the organs that he or she donated to the serial killer? Will "part" of his soul go to heaven, and the "other part" remains inside the killer's body?

Christ promoted equality and fairness, but all his disciples were XY chromosome-bearing humans. No women. Why? Was Jesus gay, mysogynous or a male chauvinist? The Bible does not have a clear answer to this topic. Did Jesus Christ and his disciples know of the existence of the American and Asian continents and of the people who lived there?

Will those fraudulent televangelists (there are lots of them), who "spread" the word of God and make confessions in front of their followers and television viewers while sobbing and crying to be forgiven, go to heaven?

Will those politicians who claim they don't hate anybody because they are Catholics and the same politicians who have knowledge of inside trading in the stock market and own several mansions go to heaven or hell? How about those trial lawyers who try to win cases by "twisting" evidences to favor their clients, and those businessmen who make obscene profits from hoarding and price-gouging, will they to go to heaven after making confessions?

Americans living in the south (also known as the Bible belt) are well known as deeply religious Christians. During the late 70's when I was offered a fellowship in neuromuscular diseases, I had the occasion to look for an apartment close to the medical school. After contacting several housing managements by phone while staying in a motel, I found one that informed me that there was a vacancy available for a 2-bedroom apartment. I travelled 10 miles by taxi to discuss the rental policies. While waiting at the lobby for the manager and while I was staring at the pictures of televangelists and churches posted on the office wall, I overheard one employee telling the other employee with a distinct southern drawl, "I am sorry, I thought he was white." To cut the story short, I was told there was no vacancy available in that apartment compound. In essence, a person can be religious and, at the same time, racially discriminatory. Praise the Christian Lord of the Bible belt!!

Thou shalt not kill. I find this commandment quite difficult to understand. Will those religious soldiers who went to war to serve their country and kill hundreds of enemies go to hell? Or will they be rewarded and go to heaven? Do the predators in Serengeti, such as wild dogs, hyenas, lions, cheetahs, leopards, and a lot more, have souls? If they do, do they go to hell? How about bacteria, fungi, and viruses that kill thousands and thousands of humans and animals every year? Believers are told that all things in this universe are God's creation, while he/she/it allows them to kill one another? If a person with Alzheimer disease who was very religious, generous, and benevolent prior to onset of dementia suddenly became paranoid and violent (some people with this disease or other forms of dementia may become violent, depressed, paranoid, and confused) and harmed or killed his relatives, will he/she go to hell? What will God do to those scientists who sacrifice laboratory animals after subjecting them to drug and surgical experimentations? From what I was told, these animals are God's creations and, just like humans, they have rights to live and enjoy life in this world.

Atrocities and victims of the Christian faith are simply too many to discuss in great detail. Here's a partial list of atrocities (stellarhousepublishing.com/victim):

1. Between 315 CE and the 6th century, thousands of Pagans and priests were slain. Children were also executed upon the order of a Christian emperor named Theodosius (408–450).

2. A world-famous female philosopher, Hypatia of Alexandria, was torn to pieces by a Christian mob led by a minister named Peter in 415.

3. In 782 CE, Emperor Charlemagne ordered the beheading of 4500 Saxons when they refused to convert to Christianity.

4. Battle of Belgrad (1456): 80,000 Turks were slaughtered on the command of Pope Urban II.

5. In the 16th and 17th century, while English troops were trying to civilize Ireland, thousands of Gaelic Irish were victims of carnage.

6. During the Crusades (1095–1291), millions of nonbelievers were slaughtered.

7. From 385 CE until 1600, thousands of heretics were beheaded.

8. From the beginning of Christianity to 1484, several thousand witches were burned at the stake or hung.

   Religious wars in the 15th century, which included crusades against the Hussites and apostates in England, killed thousands as a result of the order of Pope Paul III. In the 17th century, in a war between Catholics and Protestants, 40% of the population in Germany was decimated

9. Crimes committed against Jews are indescribable. Jews were enslaved, their communities sacked, and 10,000 were killed in

Poland in 1290. Hundreds were killed in France and Germany. In 1648, in the Chmielnitzki massacres (Poland), 200,000 Jews were slain.

10. During the years of Spanish conquistadores and Columbus' journey to the Americas, thousands and millions of natives were killed in the name of Christianity (during the period 1500–1900).

11. On November 20, 1864, Colonel John Chivington, a former Methodist minister, gunned down a Cheyenne village of 600, mostly women and children.

12. Everybody knows about the millions of Jews killed by Hitler, a Catholic, in WW II, but there were also thousands killed in the years 1942 and 1943 in Croatian exterminations camp run by Catholic Ustasha under the dictator Ante Pavelli, a Catholic. Many of the killers were Franciscan friars. The Pope knew about the killings, but did nothing to prevent them.

13. In 1954, the US deprived the Buddhists in Vietnam of much-needed aid. Food, as well as technical and general assistance, were given to Catholic Vietnamese.

14. During the Rwanda Massacres in 1994, ethnic wars erupted throughout the country. Catholic priests and nuns were suspected of actively participating in the massacres of non-Christian ethnic groups.

Any discussion about the atrocities of the Christian faith must include the horrors of the Spanish Inquisition. It was established in 1478 by the Catholic Monarchs King Ferdinand II of Aragon and Queen Isabella I of Castile. It was intended to identify heretics and to force them to convert to Christianity. It was reported that 150,000 people were prosecuted. In addition, between 3,000 and 5,000 people were executed, most of which were Jews and Muslims.

"The perpetrators of the inquisition – the torturers, informers and those who commanded their actions were ecclesiastics of one

rank or another. They were men of God – popes, bishops, and priests."
– *Sam Harris*

Such are the paradoxes in religion—"Be cruel and kill in order to be kind and survive." (Paradox Definition and Meaning, www.dictionary.com).

In this modern times, cruelty in the name of religion persists. In the March 29, 2022 issue of the Daily Gazette of Schenectady, New York, an article titled, "Indigenous tell pope of abuses endured at Canada School" was published. It was reported that more than 150,000 tribal children were isolated from their native culture and forced to attend residential schools run by Catholic missionaries. They were beaten for speaking their native languages and physically and sexually abused by the clergies from the 19th century until the 1970's. Such rampant abuses were recently acknowledged by the Canadian government. The pope later apologized to indigenous people of Canada. Praise the Lord. ❦

CHAPTER 14

Antarctica and the Arctic, pyramids, the Bering Land Bridge, the Mayans, lost civilizations, sunken cities, star clusters, and megaliths

> "A millennium before Europeans were willing to divest themselves of the Biblical idea that the world was a few thousand years old, the Mayans were thinking of millions and the Hindus billions"
> – *Carl Sagan, Cosmos*
>
> "When you understand why you don't believe in other people's gods, you will understand why I don't believe in yours."
> – *Albert Einstein*

The more you know, the more you know you ought to know. According to Aristotle, "The more you know, the more you know you don't know." Either way, knowledge is power. It is almost impossible for someone with knowledge based on facts to argue intelligently with someone whose knowledge is based on a belief system or hearsay.

The literature on ancient relics, megaliths, and the remains of old and lost civilizations that predate the Bible is voluminous. A wealth of information lies hidden and submerged in various bodies of water in planet earth. Unsurprisingly, none of these are discussed in the Bible. From what I could discern from e-mail exchanges in our group, none of my classmates are familiar with authors like Graham Hancock, Robert Temple, Robert Bauval, Adrian Gilbert,

Paul Von Ward, David Hatcher Childress, Will Hart, Robert Steven Thomas, and Michael Tellinger, among several others, who had done research on archeology, the origins of human species, ancient gods, star systems, pyramids in different parts of the world, sciences of the ancients, the missing link in the history of evolution, and Sumerian history. It is possible that some of them may have knowledge about the subjects, but they turn a blind eye because of the concern that they might shatter their beliefs. One classmate, long retired and got pissed off with me, who browses the internet several hours a day, is aware of the existence of unsolved mysteries in the past but elected to ignore them for the fear of being isolated from their religious group. Another classmate, an internist, who lives and practices in the Old Dominion state expressed his blatant ignorance and criticized me for reading too many "fiction" books. This is the nature of my beloved classmates' mindset, despite their erudite scientific mind.

"I don't understand why so many people who are sophisticated in science go on believing in God. I wish I did." – *Richard Dawkins*

The arctic and Antarctica are two contrasting poles: the former has no solid land underneath the ice, while the latter has a solid land mass covered by ice. According to the Physics Fact book, the thickness of the ice cap of Antarctica ranges from 2000 m (4,479 feet) to 4,470 m (15,650 feet. In 1953, the late Charles Hapgood, who taught the history of science at Keene College, New Hampshire, theorized that Antarctica was not always covered with ice and was located outside the Antarctic circle, where the temperature was warmer. The continent moved to the southern circle due to "southward earth-crust displacement" around 4000 BC, at the time when Sumerian civilization, along with sky gods, began to appear in the Fertile Crescent of the Middle East. The theory proposed by Hapgood was based on Piri Reis, an Ottoman-Turkish Navy admiral and cartographer who drew a world map in 1513. The map showed the coastlines of Europe, Africa, Brazil, and possibly Japan, with reasonable accuracy along with the ice-free coastlines of Antarctica. Who discovered Antarctica is a matter of debate, and what the sources of the Piri Reis map were remains an unsettled issue. However, it is clear that the map was drawn at least 300 years before the discovery of the frozen continent.

Who built the pyramids and the Sphinx in Egypt? When and why, and what do they symbolize or represent? Anybody can speculate, but the Bible cannot, since it is a mind-controlling compilation of many books written and revised many times by different authors without the use of science. When Graham Hancock (formerly the East Africa correspondent for The Economist and a correspondent for the London Sunday Times) asked John West (1932–2018), a geologist and an American lecturer, about why Egyptians have been so unwilling to consider that the Sphinx might have been built by some other advanced civilization, the answer was, "The reason, I think, is that they're quite fixed in their ideas about the linear evolution of civilization. They find it hard to come to terms with the notion that there might have been people, more than 12,000 years ago, who were more sophisticated than we are today." West and Robert Schoch (a Boston University geologist) delivered a bombshell report to the archeological community stating that the Sphinx showed evidence of rainfall erosion, which occurred during the rains that fell during the transition period of northern Africa from the last Ice Age to the interglacial epoch, 10,000–5000 BC.

"On the search for truth—that everything in nature seems to hide—man needs the assistance of all his faculties. All the senses should be awake." – *Robert G. Ingersoll*

In the search for truth, one must use reason. Through the years, man has been told the truth, but has also been misled and misinformed. Even in this modern era, far-left and far-right media and politicians have narratives based on bias and ideology. Since the search for documents pertaining to the distant past is almost always difficult, one must rely on the use of all mental faculties and common sense and set aside fixated beliefs and arrogance to promote progress.

How and who built the pyramids is a mystery. According to respected and erudite experts, slaves did not build them. The Great

Pyramid is a huge structure, weighing approximately six million tons and occupying 13 acres. It stretches to more than 750 feet along each side. It is 481 feet tall. More than 2.5 million blocks of stone, each weighing several tons, were used in the construction. The polar radius of the earth is equivalent to the height of the pyramid multiplied by 43,200. The equatorial circumference of the earth is equivalent to the measurement of the base perimeter of the pyramid multiplied by 43,200 (Graham Hancock, author of the Fingerprints of the Gods). Such mathematical precision, if truly accomplished by ancient Egyptians, not to mention the apparently crude techniques and materials they used to lift tons of stone blocks and align and secure them with accuracy, is beyond comprehension. To make things more mysterious, it was theorized that the pyramids were built to align with the Orion, a bright cluster of stars conspicuously visible in the celestial equator. It is named after Orion, a Greek hunter in Greek Mythology. This alignment was believed to be associated with precession, which is caused by the very slow motion of the earth, like a wobble that takes about 26,000 years to complete a full cycle. The half of this cycle is 13,000 years, almost the same duration in which the pyramids were built (12,500 BC). However, traditional Egyptian researchers date back the construction to 2550–2490 BC. Who do you believe? And who is correct?

Although the pyramids have always been associated with ancient Egyptians, it should be noted that they have also been found in the jungles of Mexico and Guatemala. They were presumably constructed by the Mayans. There are also pyramids in China, the largest of which is located in Xian, a central city in China. Unfortunately, for various reasons, archeologists and geologists could not provide the world public with any further details about them.

The Bering Land Bridge or Bering Strait connected eastern Siberia to the US before the biblical great flood. The bridge existed during the Pleistocene epoch, 700,000 years ago, when the sea level dropped as earth's water and precipitation became frozen in large continental ice sheets and glaciers. Soon, the bridge became a dry land, enabling humans to migrate. Plants and animals of similar species exist on both continents. Now, we know why Eskimos and Asians (and some central Americans) appear phenotypically similar. The Bible states that God created man in his image; however, the book of Genesis says, "Let us create man in our image." So the "singular God" became "plural," which would be the equivalent of polytheists in Sumerian civilization. Who created the Asians, including Filipinos? The apostles who went on and spread the words of God, how far did they go? Were they aware of other lands like the Americas and Southeast Asia? If somebody tells me that Columbus and Magellan were descendants of the Apostles, then they don't impress me as kind and benevolent because they murdered thousands of natives in the land they "discovered," particularly the Philippines and America, while they spread Christianity.

Spaniards began to conquer parts of Mesoamerica during the 1520s. The Mayan existence began approximately during the first millennium BC and declined during 830–900 AD. During the conquest of the Mayan Empire, numerous historical documents and cultural aspects were destroyed because they were considered a Pagan civilization. Social life was disrupted. The destruction was similar to the destruction of 2nd century Jewish books and that of the Talmud by the Roman Catholic church. Ancient Mayans were very intriguing and interesting people. According to the legend, they were visited by extraterrestrial beings led by Quetzalcoatl, a feathered serpent who taught them astronomy, medicine, mathematics, and agriculture. He was considered their spiritual leader and was associated with the creation of mankind. Signs that indicate that the Mayans were visited and educated by extraterrestrials were artifacts and statues of beings operating the inside compartment of flying machines. They probably originated from Pleiades, a cluster of stars in the constellation of Taurus. It can be said that if Quetzalcoatl is to the Mayans, Oannes is to the Babylonians, a fishtailed amphibious being who founded civilization on Earth. Oannes is depicted in the stone relief of the Assyrians as a half-man half-fish who came from the Sirius star system.

Seventy percent of the Earth's surface is covered by water, and yet a very small percentage has been explored by man. Legends, tales, and myths are abound, but man has now learned that some of them have substantive scientific merit despite the intentional destruction of some documents by fanatics and religious zealots. Adherence to traditional beliefs and resistance to acquiring new knowledge slowed down the progress of science since the establishment of organized religion.

In modern times, however, scientists are now discovering various relics and remains of ancient civilizations submerged deep in the ocean all over the world. The enigmatic and legendary Atlantis continues to tickle the cerebral hemispheres of freethinkers and various scientists. Now, according to David Hatcher Childress, there is the possibility that there was once a continent now submerged in the vast Pacific Ocean, called Lemuria (Pacifica by geologists and Mu or Pan by mystics). It is not surprising that these continents might have existed in the remote past and disappeared from the surface of the earth following earth-crust displacement or massive geologic and tectonic cataclysmic events.

The global Biblical flood (described in the Epic of Gilgamesh and Noah's flood 10,500 BC), according to the preponderance of geologic evidence, did really occur. The globe was inundated by the flood as ice began to melt, resulting in the sinking of ancient coastal cities that included the Egyptian Mediterranean city of Alexandria, antediluvian cities of Sumer, Yonaguni Island in Japan, and the city of Khambhat in India. Around this period, the Bering Land Bridge became submerged. Robotic exploration conducted in sea along the Bahamas, Florida, and Cuba revealed evidence of submerged relics of ancient cities. The rain that fell down might have caused the erosion found in the Sphinx. Most mainstream archeologists, by virtue of their egos, dispute these findings.

Megaliths are archeological wonders. They are found in many parts of the world. One particular megalith is found in the Salisbury Plain in Wiltshire, England. It is an iconic archeological wonder and is quite enigmatic. It has an outer ring of vertical sarsen standing stones, each around 13 feet high, 7 feet wide, and each weighing 25 tons. Inside is a ring of smaller blue stones. Archeologists believe that it was constructed between 3000 BC and 2000 BC. Several other megaliths weighing several tons are found in France, Bosnia, Costa Rica, and Indonesia. The giant heads of Easter Island in the southern Pacific Ocean are equally intriguing. They are 13 feet tall (head to torso), each weighing 14–80 tons. They were built and transported for several miles and erected between 1400 and 1600 AD. They are all inland-facing and are called "Mo'ai" by the natives living in eastern Polynesia; "Mo'ai" means "statue" in Rapa Nui, an Eastern Polynesian language. There are approximately 800 statues. They were considered powerful deities by the natives. Who built them, and why were they built? Why are all of them inland-facing?

It is quite obvious that the men (the history did not mention women and everything is interpreted through male eyes) who wrote the Bible did not have any idea about the existence of archeological wonders and relics that are found all over the world. Realistically, those who wrote the Bible never travelled outside the Mideast and none of them knew nothing about science and nature. Everything was about control of the mind and emotion which only substantiate the fallibility of the Bible, a document written and restructured by men. ❧

CHAPTER 15

The origin of the human species

> "Darwin's theory of evolution is a framework by which we understand the diversity of life on Earth. But there is no equation sitting there in Darwin's "Origin of Species" that you apply and say, "What is this species going to look like in 100 years or 1000 years? Biology isn't there yet with that kind of predictive precision."
> *– Neil deGrasse Tyson, American astrophysicist and science communicator*
>
> "Biological evolution is too slow for the human species. Over the next few decades, it's going to be left in the dust."
> *– Ray Kurzweil*

In his book titled, "Origins of Existence," Fred Adams, professor of physics at the University of Michigan, defined life as anything that can replicate itself and sustain itself with metabolism; the former is mediated by the genetic materials in the chromosome located inside the nucleus of a cell, and the latter is mediated by the mitochondria located in the cytoplasm. For life to continue, the two systems must work together as a unit and, in the grand scheme of things, are governed by the laws of physics in concert with chemical reactions and the biologic process. Following the laws of physics, the same mechanism is applicable to any form of life in this vast universe.

Life on planet earth began with the appearance of single-cell organisms, such as bacteria, which reproduced asexually, meaning reproduction without parent cells. Soon, complex multicellular animal forms appeared on Earth, each with specialized pluripotential cells—the spermatozoon and the ovum. The union between the spermatozoon (male cell) and the ovum (female cell) is the beginning of a complex series of cellular proliferation, migration, and exchange of genetic materials between a man and a woman. Soon, three germ cell layers—the ectoderm, mesoderm, and endoderm—are formed, from which various organs are derived. The endoderm gives rise to the respiratory tracts, the urinary bladder, the thyroid and parathyroid, the liver and pancreas, tonsils, the thymus gland, and the epithelia of the tympanic cavity. The striated and nonstriated muscles, cartilage and bone, connective tissues, dura, parietal peritoneum and pleura, endocardium and myocardium, blood and lymph vessels, gonads, spleen, and adrenal gland are derived from the mesoderm. The ectoderm gives rise to the brain and spinal cord, skin, sensory epithelia of the eyes, ears and nose, mammary glands, pituitary glands, enamel, sinuses, and oral cavities. How a soul or spirit is formed within a multicellular human being and is separated at the time of death is nonsensical and farcical, and way beyond comprehension.

The universe consists of countless galaxies, each one composed of billions of stars together with planets that revolve around each one of them. There are also countless celestial objects, such as moons, that orbit planets, asteroids, meteors and meteorites, and comets. The distance between galaxies and stars is measured in thousands or millions of light years (a light year is equivalent to the distance travelled by light in one year, approximately 5.8 trillion miles, at a speed of 186,000 miles for a second). Following the "Big Bang", the galaxies continue to move away from each other. It is more than likely that there exist other civilizations outside of our planet Earth that are either more advanced or less sophisticated than us, or they have been in existence for thousands of years prior to the birth of our solar system. If this is the case, do those civilizations have Jesus Christ, the

Holy Trinity, the Virgin Mary, etc., in their biblical history, if there is any? I seriously doubt it.

The complexity of the formation of another human being following the union between two microscopic parent cells, and the colossal formation of the unimaginable and immeasurable dimension of the universe, are simplistically, using a magic wand, condensed and squeezed in 6-7 days by the biblical account of creation. In a child with a very young central nervous system, the biblical teaching of creation can be easily inculcated into their gullible and vulnerable minds through daily and repeated teachings, at home and in school. These are reinforced by the threat of eternal punishment in hell for disobedience. The knowledge and the threat that were successfully implanted into their brains, unless damaged by significant brain pathologies such as stroke, trauma, dementia, and infectious diseases, become mystifyingly permanent. What is so intriguing is how hostile believers become when their beliefs are challenged by nonbelievers.

For thousands of years, after the disappearance of the Neanderthals, a thinking or wise man known as Homo sapiens, to which modern man belongs, existed and roamed the earth. They were hunter-gatherers and lived by foraging wild plants and hunting wild animals. Out of clear blue or black sky, around 6000–3000 BCE, complex civilizations suddenly sprouted in many parts of the world. All of them had accounts of creation in which gods came down from heaven and created man in their image. These civilizations were established in Egypt, Sumer, Peru, Mexico, China, and the Indus Valley. They built pyramids and megaliths, but why and how they were constructed remains a mystery. According to Christopher Dunn, a master craftsman and engineer, these pyramids were built to serve as power plants, and the large pyramid was built as a giant acoustical device to provide harmonic resonance with the earth and convert the earth's vibrational energies to microwave radiation. These civilizations had keen knowledge of astronomy, architecture, mathematics, agriculture, and carpentry. Alternative theorists believe that these pyramids were built using advanced technology of extraterrestrial origin. They were powerful gods who influenced our history, manipulated our DNA, and shaped our cultures. Paul von Ward, an independent scholar ordained as a Baptist minister and trained as a psychologist, remarked that the time has come to resolve the ideological divisions between spiritualists and materialists and to re-examine the true meaning of "divine revelations."

The microscopic materials that hold the key to the origin and propagation of life are located in the nucleus of a cell. Inside the nucleus are rod-shaped organelles, called chromosomes, that stain deeply with histological dyes. The human genome is the complete set of full DNA content assembled and contained in 46 chromosomes in 23 pairs, 22 autosomes, and a pair of sex chromosomes. The autosomes determine the trait of the individual, and the sex chromosomes which consists of XX and XY chromosomes, determine the gender of the individual. Each chromosome has segments known as deoxyribonucleic acid or DNA, where genes, the basis of heredity, are formed. There are approximately 20,000–25,000 genes in each chromosome in humans.

The nucleic acids DNA and RNA (ribonucleic acid) are composed of three types of units: a five-carbon sugar deoxyribose in DNA and ribose RNA, a nitrogen-containing base (purines and pyrimidines), and a phosphate. DNA has two purine bases (adenine and guanine) and two pyrimidines (thymine and cytosine). The DNA molecule, or the "Watson-Crick double helix," carries the biochemical basis of heredity. The helix has polynucleotide chains held together by hydrogen bonds between pairs of bases—adenine (A) paired with thymine (T) and guanine (G) paired with cytosine (C). Genetic information is stored in the DNA molecule by means of code, where an infinite number of variations is possible.

Any discussion about DNA must include mitochondrial DNA. These genetic materials are located in the mitochondria and in the cytoplasm of eukaryotic cells. They contain genes essential for the synthesis of adenosine triphosphate, the cell's main energy source.

This DNA is inherited exclusively via the maternal line. Organs affected by the disorders of the mitochondria include the brain, muscles, nerves, eyes, ears, heart, liver, and kidneys. In other words, mitochondrial disorder is a multi-system disease. If the muscle is primarily affected, locomotion will be severely impaired owing to severe weakness and wasting of the muscles of the extremities in addition to various disabling symptoms referrable to several malfunctioning organs.

Disorders of the mitochondria, among many other genetic disorders, have been recognized in the past several decades. Diseases that used to be considered degenerative are now known to be causatively related to the presence of abnormal genes. Such discovery is made possible by progress in science, but not by daily prayers and religious rituals. Soon, a therapeutic regimen will be made available through genetic engineering and, hopefully, without religious interference.

The DNA helix resembles the Sumerian double helix snake god, the Caduceus, and the Rod of Asclepius. Asclepius is the god of medicine in Greek mythology. The rod has a single snake without wings, while the Caduceus is a winged staff with two snakes wrapped around the rod. They are were both used as a symbol of medicine. Did the Sumerians leave a message for future generations of humans to decipher and reveal to them that the basis of creation was all about DNA? How about ancient Babylonian literature depicting the reptilian origin of humanity? How about the feathered serpent of the Mayan Empire (Quetzalcoatl)? ❧

CHAPTER 16

Catastrophes, natural disasters, wars, and famine

"You might think that, by now, people would have become accustomed to the idea of natural catastrophes. We live on a planet that is still cooling and that has fissures and faults in its crust; this much is accepted even by those who think that the globe is only six thousand years old, as well as by those who believe that the earth was "designed" to be this way. Even in such cases, it is to be expected that earthquakes will occur and that, if they occur under the seabed, tidal waves will occur also. Yet two sorts of error are still absolutely commonplace. The first of these is the idiotic belief that seismic events are somehow "timed" to express will of God. Thus reasoning back from the effect, people will seriously attempt to guess what sin or which profanity led to the verdict of the tectonic plates. The second error, common even among humanists, is to borrow the same fallacy for satirical purposes and to employ it to disprove a benign deity."
– *Christopher Hitchens*

"Either God can do nothing to stop catastrophes like this, or he doesn't care to, or he doesn't exist. God is either impotent, evil, or imaginary. Take your pick, and choose wisely"
– *Sam Harris*

Everything in the universe has been in constant motion since its birth, 13.7 billion years ago. Electrons orbit the nucleus (composed of neutrons and protons) at a speed of 2200 kilometers per second. Earth orbits the sun at a speed of 110,000 kilometers per hour, 365 days a year. It rotates on its own axis 24 hours a day. Collisions among celestial objects in this universe occur on a regular basis. Asteroids, comets, and meteorites are well known to collide with all planets, including their moons. The demise of dinosaurs is believed to be due to a collision between the earth and an asteroid, stretching 6–9 miles across, roughly the size of Manhattan. When measured in human time (hours, minutes, and seconds), the solar system and the universe are seemingly peaceful, but the universe is a volatile and violent place when time is measured in astronomical terms. Besides collision among celestial objects, the gravitational tug of war among planets and moons, earthquakes, volcanic eruptions, hurricanes, solar flares, and supernova explosions, among many others, occur endlessly on a regular or cyclical basis.

Here on earth, murders, homicides, and deaths from illness, accidents, or natural disasters, to name a few, occur several times a day; 11,000 years ago, the angry, vindictive, and temperamental God of the Old Testament reportedly killed many people by drowning them in a global flood as a punishment for their sins and wickedness. Whether the flood was local or global remains undetermined, but old records showed that such an event was also recorded in ancient Chinese and Indian literature.

Although no one is certain, approximately 50–70 million people died during World War II, and it was the bloodiest conflict in human history. Among the dead were millions of Jews killed by Hitler, a Catholic. Casualties in World War I vary greatly, and approximately 9–15 million people died. The death tolls of the Iran-Iraq war of 1980–1988 were 500,000 dead Iraqis and 750,000 dead Iranians (kurzman.unc.edu/deathtolls-of-the-iran-iraq-war).

Here's the list of the top 10 worst disasters of the world (Jessica Karpilo, "10 World's Worst Disasters". Thoughtco.com/world-worst-disasters 1434989), from lowest to highest death toll.

10. Aleppo Earthquake (Syria, 1138) – 230,000 dead

9. Indian Ocean Earthquake/Tsunami (Indian Ocean, 2004) – 230,000 dead

8. Haiyun Earthquake (China, 1920) – 240,000 dead

7. Tangshan Earthquake (China, 1976) – 240,000 dead

6. Antioch Earthquake (Syria and Turkey, 526) – 250,000 dead

5. India Cyclone (India, 1839) – 830,000 dead

4. Shaanxi Earthquake (China, 1556) – 500,000–1,000,000 dead

3. Bhola Cyclone (Bangladesh, 1970) – 500,000–1,000,000 dead

2. Yellow River Flood (China, 1887) – 900,000–2,000,000 dead

1. Yellow River Flood (China, 1931) – 1,000,000–4,000,000 dead. The flood lasted for 3 Months (June – August)

On December 26, 2004, thousands of people died in the tsunami across the Indian Ocean, with waves travelling at 500 MPH. The collective death toll in several countries was 230,000 (Indonesia, Sri Lanka, India, Maldives, and Thailand). The several tornadoes that swept across eight states in the US, from the "Bible states" to the Midwest, in early December 2021, were devastating. Mayfield, Kentucky, was hit the hardest. Shortly after, a super typhoon hit the central islands of the Philippines, killed at least 370 people, and left thousands injured. Property damage was extensive.

Here's the list of the 10 deadliest natural disasters in the Philippines, from lowest to highest death toll (Newsinfo.inquirer.net (524569))

10. Typhoon Trix, October 16, 1952, Bicol – 995 killed

9. Typhoon Sendong, December 16, 2011, Northern Mindanao – 1080 killed

8. Landslide, February 17, 2006, Guinsaugon, Southern Leyte – 1,126 killed

7. Mayon Volcano, February 1, 1814, Albay – 1,200 killed

6. Taal Volcano, January 30, 1911, Luzon – 1,300 killed

5. Typhoon Ike, August 31, 1984, Siargao – 1,363 killed

4. Magnitude 7.8 earthquake, July 16, 1990, Baguio City – 1,621 killed

3. Typhoon Pablo, December 3, 2012, Southern Island of Mindanao – 1900 killed

2. Tropical storm Uring, November 15, 1991, Ormoc, Leyte – 5,100 killed

1. Tsunami and earthquake, August 16, 1976, Moro Gulf, Mindanao – 5000–8000 killed. Deadliest tsunami in Philippine history

A severe scarcity of food is called famine. Its causative factors include war, natural disasters, farming failure, population imbalance, economic failure, governmental policies, and the widening gap between affluence and poverty. During the 20th century, an estimated 7–120 million people died from famines worldwide. Africa has been the most affected since 2010. The World Food Programme warned that 45 million were on the brink of famine across 43 countries. Afghanistan is on the brink of collapse following the US' withdrawal from the country. If the COVID pandemic becomes protracted, the risk and incidence of famine are likely to increase, and poor countries are likely to suffer more.

The question often asked by nonbelievers, skeptics, agnostics, and atheists is, "Why does God allow calamities to happen? Why did the Christian God allow the destruction of the World Trade Center in New York City by the followers of Islam? The believers' answer is always, "It's God's will, and He loves us." It is somewhat ludicrous and unthinkable that God's will for so many years should be that way. There must be a better explanation. Praise the Lord. ✑

CHAPTER 17

Politics in religion

"Those who think religion has nothing to do with politics understand neither religion nor politics.... The things that will destroy us are: politics without principles, pleasure without conscience, knowledge without character, business without morality."
 – *Mahatma Gandhi*

"I never considered a difference of opinion in politics, in religion, in philosophy, as cause for withdrawing from a friend."
 – *Thomas Jefferson*

"Religion and government will both exist in purity, the less they are mixed."
 – *James Madison (1751–1836), 4th US President, statesman, diplomat and philosopher*

"Nothing could be more dangerous to the existence of the republic than to introduce religion into politics"
 – *Robert G. Ingersoll*

"Leave the matter of religion to the family altar, the church, and the private schools, supported entirely by private contributions. Keep the church and the state forever separated."
 – *Ulysses S. Grant (1822–1885), Union General, and 18th President of the U.S.*

It's been said that religion and science do not mix, or, as they say in chemistry, are not miscible. This is understandable, since the basic foundation of the former is belief and faith, while the latter is based on observation and reason. To some, religion should not be mixed with politics, because it will inevitably affect the responsibilities of their leaders. However, the molecular binding between politics and religion in most countries is so tight that one cannot thrive or exist without the other. Indeed, in countries, such as the Philippines, religion is tightly intertwined with politics. They are so inseparable that a person becomes unelectable if he/she is an agnostic or an atheist. Similarly, in a predominantly Islamic country, it is unimaginable for a Christian or a Jew to be elected to any government office. This is the nature of religion, an institution that is so venerable and sanctified in paper but is disreputable, unconsecrated, and pathetically corruptible in practice, particularly when mixed with politics. Either one could weaponize the other to intimidate opponents with different beliefs or ideologies. What is so appalling is what politics can do to a religious and honest person when he or she runs for office.

During the Italian renaissance, the church was quite a dominant power. It controlled politics, kings and very wealthy. Pope Leo X (1475-1521), who came from the most influential families in Europe, was famous for abuses of power. The exploitation of people, excessive spending, selling the forgiveness of sins, and corruption of religious practices were the root causes of Protestant Reformation led by Martin Luther, who was later excommunicated by the church in 1521. At a lavish Good Friday gathering in the Vatican, he made an infamous and damaging quote about Christianity. While raising a chalice of wine he said, "How well we know what a profitable superstition this fable of Christ has been for us and our predecessors." This quote generated controversies in the Christian community, and many thought the quote was a mistranslation. Others thought otherwise.

Back in the mid-80s, a high-ranking priest in the Philippines led the effort to revolt against the abuses and growing dictatorial power of a popular president. He made an urgent call to the citizens to take to the street as a show of force, famously called the People Power Revolution. Eventually, that president was kicked out of the office. Such is the power of religion. In the atheist People's Republic of China, however, the powerful ruling party can easily quell any religious uprising; in fact, under its current leader, there is a strong push to sinicize all religions in the country to conform to government principles. Religious organizations are required to accept the ideology of the communist government or their members will be tortured, sent to prison, or killed, which they deny. Despite pressure from the international community, the ruling party's tenacious adherence to government doctrines remains firm. They continue to crack down on Muslim groups in northwest China for allegedly holding separatist ideas.

In some Islamic countries, religion has a strong influence on managing government affairs. In Iran, direct rule by clerics and neofundamentalists has been institutionalized, and their ideology embodied in the constitution. In essence, religion controls almost all aspects of life, including the economy and the military, in this country.

For most of the 20th century, organized religions were repressed by the former Soviet Union. In the 80s, Mikhail Gorbachev, the 8th and the last leader of the Soviet Union and a Christian, declared a policy of glasnost ("openness"), which allowed tolerance for the practice of religion. As in China, religious activities in Russia remain under the control of the communist party. The Russian Orthodox Church is a major element of Russian culture, while Muslims constitute Russia's second largest religious group.

According to the Pew Research Center, in the US, Christians (Protestants, Catholics, Mormons, Jehovah's Witnesses) comprise 70% of the religious groups; the others are from non-Christian faiths (Jewish, Muslims, Buddhists, and Hindus; 5.9% overall) and unaffiliated groups (atheists and agnostics, 3.1% and 4.0%, respectively, and nothing in particular, 15.8%; 22.8% overall). The 2008 ARIS survey showed that, in the US, belief in God is the lowest in the West at 59% and highest in the South at 86%. Free exercise of religion in the US is guaranteed by the First Amendment to the country's Constitution, which is in sharp contrast to other countries. This prevents the government from having any authority in religion, but the relationship between politics and religion in the country is quite complex. Some Americans are willing to set aside their religious affiliation in favor of political preference, especially liberal-progressives who view religion as an extension of conservative ideology, and when conservative views are opposed, the other party would cry foul for attacking their religious freedom. The choice of political candidate also depends on economic issues, where candidates who are business friendly will get the vote.

To add to the complexity is the subject of abortion, one of the most controversial and very difficult and contentious issues that divide American society. According to the Pew Research Center, Jehovah's Witnesses, Mormons, the African Methodist Episcopal Church, Assemblies of God, the Catholic Church, the Church of Jesus Christ of Latter-day Saints, Hinduism, the Lutheran Church Missouri Synod, and the Southern Baptist Convention oppose abortion rights, with few or no exceptions. The Episcopal Church, the Evangelical Lutheran Church in America, and the United Methodist Church support abortion rights with some limits. Conservative Judaism, the Presbyterian Church (USA), Reform Judaism, Unitarian Universalist, and the United Church of Christ support abortion rights, with few or no limits. Islam, Buddhism, the National Baptist Convention, and Orthodox Judaism have no position on this matter.

Religion is supposed to unite people, but when mixed with politics, relationships among friends, neighbors, co-workers, and family members break down, and it can be toxic. Disinformation, mudslinging, filthy criticisms, personal attacks, and fake accusations are ridiculously routine. In one radio talk show program, the host told his audience during the 2016 presidential campaign that a vote for a Republican candidate was "immoral." In my clinical practice, I frequently encountered some patients who were staff members in a local church and who openly uttered the same thing without hesitation. Similarly, priests, pastors, and televangelists had taken sides and convinced their congregants to vote for the candidates their church endorses, going to the extent of telling them that a vote for the other candidate can be sinful. One can imagine who that deeply religious and loyal congregant will vote for.

Religion is faith, and politics is power, and they have been this way since humans were taught to become civilized in order to establish a system of government and to revere the supreme gods that came down from the sky. So now we know that religion and politics were born together at the same time thousands of years ago. ❧

CHAPTER 18

Religion in medicine

"Theology still tries to interfere in medicine where moral issues are supposed to be specially involved, yet over most of the field the battle for the scientific independence of medicine has been won. No one now thinks it impious to avoid pestilences and epidemics by sanitation and hygiene; and though some still maintain that diseases are sent by God, they do not argue that it is therefore impious to try to avoid them. The consequent improvement in health and increase of longevity is one of the most remarkable and admirable characteristics of our age. Even if science had done nothing else for human happiness, it would deserve our gratitude on this account. Those who believe in the utility of theological creeds would have difficulty in pointing to any comparable advantage that they have conferred upon the human race."
 – *Bertrand Russell, author of Religion and Science*

"As a scientist, I am hostile to fundamentalist religion because it actively debauches the scientific enterprise. It teaches not to change our minds, and not to want to know exciting things that are available to be known. It subverts science and saps the intellect."
 – *Richard Dawkins*

"A man's ethical behavior should be based effectually on sympathy, education and social ties and needs; no religious basis is necessary. Man would indeed

be in a poor way if he had to be restrained by fear of
punishment and hope of reward after death."
– *Albert Einstein*

Medicine is one branch of science that, when mixed with religion,
may be considered compatible or even a necessity and spiritually
comforting to some. To others, incorporating religion into medicine
is anachronistic and incongruous. How often do we see parents
praying for the rapid recovery of their child after an injury or
while enroute to the hospital emergency room for treatment? In
the past, clergies and various religious leaders were called upon to
perform healing or exorcism when somebody was possessed by the
devil. Some maintain that religion has no place in medicine since
it will certainly not alter the course of incurable diseases, such as
metastatic cancer, Alzheimer's disease, or amyotrophic lateral
sclerosis, a relentlessly progressive neurological disorder that affects
the spinal cord and the brain. Advocates of incorporating religion
and spirituality into the practice of medicine are very much aware of
the inevitable outcome of those dreaded diseases, and they feel that
patients should die with dignity and comfort through prayers. Most
physicians have formal training designed to respect the dignity of a
dying patient and to exercise sympathy and compassion with his or
her relatives. However, strong differences become overtly displayed
when patients and/or relatives insist that prayers can substitute for
well-established and lifesaving therapeutic regimens for treatable
diseases. Here, differences in religious practices play a major role in
the decision-making process in a multi-ethnic, multi-cultural, and
multi-faith society.

Members of Jehovah's Witnesses claim that their religion is the
true religion. They do accept medical treatment, but they do not accept
blood transfusion, autologous blood, or blood products because God
forbids taking blood to sustain the body. Almost all will accept organ

transplant, and some will accept albumin or fresh frozen plasma. The American College of Surgeons Committee on Ethics stated, "In the emergency setting, patients often do not have time to make their own decisions. Sometimes patients are unconscious, or sometimes the surgeon does not have time to evaluate the patient's capacity, as this process would delay emergency treatment and lead to the patient's death. In the emergency setting, therefore, surgeons are on solid ground in treating Jehovah's witness patients with transfusions under the standard of presumed consent. Some patients will have very clear evidence directives about not accepting blood. To the degree the surgeon can be certain that the advance directive applies to the specific patient in the specific setting, then it is acceptable to follow that directive."

Brain death is another controversial issue in which religion, the law, and medicine can intersect. Neurologically, brain death is defined as the total and permanent cessation of cerebral and brainstem activities, even if the spinal cord is intact. Brain death has been legally established in all US jurisdictions as death, but it should be emphasized that it is not synonymous with coma and a persistent vegetative state. When the patient is declared dead, the life support system can be disconnected. Controversy arose when a patient was pregnant. This situation occurred in 2014, when a Texas hospital refused to discontinue the life support system in a woman who was 14 weeks pregnant. The relatives sued the state of Texas and won. The judge sided with the family and stated that the state misapplied the law and that the fetus was not viable. In other situations, even if a neurologist declares a patient brain dead, the life support system must be maintained if there is religious objection. New Jersey has legislated a complete religious exemption from brain death. A 1991 law provides that death "shall not be declared "based on neurological criteria when that would violate the patient's "personal religious beliefs." To some religious organizations (Buddhists, Japanese Shintoists, Native Americans, and Orthodox Jews), death means complete cessation of cardiopulmonary activities. Some Christians are considering asserting similar views. In this situation, prolonging the life support system of a brain-dead patient can jeopardize the viability of body organs that are being considered for transplant.

The Catholic Church has fully acknowledged developments in the science of medicine. It has not issued any serious challenge to brain death and considers brain death an indication that the patient has died. It has also been acknowledged by many Muslim scholars and organizations, including many Islamic nations, as true death.

When a patient is declared brain dead, the next order of the procedure is to discontinue the life support system. The lungs will no longer expand, but the heart will continue to beat on its own for several minutes until the oxygen supply is totally cut off. The ventricles will then fibrillate, contract irregularly and weakly, and then stop beating. In other words, the total somatic death that follows the discontinuation of the life support system can be argued as iatrogenic. Such is the argument presented by other religious groups, hence the reason for not accepting brain death as actual death. These groups prefer the natural cessation of cardiopulmonary function as the equivalent of death, even if the brain is totally dead.

Here's my recommendation to religious groups who object to brain death as being the equivalent of death. They should be invited to look at a dead brain in an artificially maintained heart and lung function when the patient becomes totally dead. A dead brain in autopsy looks macerated because of autolytic changes that occur while the brain is deprived of oxygen and blood flow despite being connected to a life support system. Any layperson will be convinced that a dead brain, which looks macerated, has zero chance of recovery, not to mention the unsuitability of other organs for organ procurement, the longer the heart and lungs are kept functioning by artificial means. ❧

CHAPTER 19

Navigating and redirecting the mind to the path of reality and common sense: The central nervous system, not the soul (if there is any), controls human mental activities – that's the reality

"The only good is knowledge and the only evil is ignorance."
– *Socrates, Greek philosopher*

"Religions are for a day. They are the clouds. Humanity is the eternal blue. Religions are the waves of the sea. These waves depend upon the force and direction of the wind—that is to say, of passion; but Humanity is the great sea. And so our religions change from day to day, and it is a blessed thing that they do. Why? Because we grow, and we are getting a little more civilized every day, and any man that is not willing to let another man express his opinion, is not a civilized man, and you know it. Any man that does not give to everybody else the rights he claims for himself, is not an honest man."
– *Robert G. Ingersoll*

"What is it the Bible teaches us? – raping, cruelty, and murder. What is it the New Testament teaches us? – to believe that the almighty committed debauchery with a woman engaged to be married, and the belief of this debauchery is called faith."
– *Thomas Paine*

"Faith: not wanting to know what is true"
– *Friedrich Nietzsche (1844-1900), German philosopher*

The power of indoctrination (or mind control) is phenomenal and astonishing, and its effect on humans is usually retained for life for as long as the gray and white matter remain intact. For it to be effective, it must be initiated during childhood when the axodendritic synapses are relatively meager and the interneuronal nerve fibers are flexible. Constant and steady neural stimulation—equivalent to teachings and inculcation of religious ideologies—will result in the formation of a pool of synapses and neuronal networks in the brain that are stored and activated under certain circumstances. Passing the baton of faith to children and the fear of eternal damnation in hell when they commit sin are the stimuli needed to sensitize these neuronal networks. This is analogous to "kindling," wherein certain parts of the brain can be sensitized when constant subthreshold electrical or chemical stimuli are applied. This process can lead to recurrent epileptic attacks, which can be triggered by several factors. Indoctrination is large-scale kindling that affects various areas of the brain that regulate and control emotions. These areas of the brain, such as the cingulate gyrus, hippocampus, amygdala, hypothalamus, and basal ganglia, are collectively known as the limbic system. These areas of the brain can be activated when a religiously indoctrinated person is challenged by someone who questions his or her religious beliefs. Obviously, these areas of the brain are not "kindled" in nonreligious people.

As discussed in Chapter 12, all parts of the central nervous system have certain functions that mediate various activities, such as locomotion, sensation, smell, vision, hearing, emotion, social behavior, and memory. The integration of all these functions, along with learning from past experiences and the ability to store and retrieve recent and remote information, determines the intellectual level and behavior of an individual. It should be emphasized that during childhood, there is basically no resistance to any kind of indoctrination, given that the child has a normal cerebral cortex. The child is simply a very vulnerable percipient. When acquired repetitively, knowledge is easily and permanently incorporated into the neural fabric of a young central nervous system. A person with a higher intellect who becomes a medical doctor or a scientist may remain a believer if the indoctrination process during childhood was strong and successful, and he/she will tenaciously adhere to their stored beliefs when challenged. If so, can it be assumed that nonbelievers were not successfully indoctrinated? How many neurologists, psychologists, astronomers, and other scientists are agnostics or atheists? Can this neural fabric be unloosened by opening a new avenue to reality and by the willingness to be intellectually open-minded and farsighted, regardless of the degree of the indoctrination process during childhood?

What is a sin? It is an immoral act or transgression of divine law. If a person was born with rogue neurons and abnormal chemical neurotransmitters in his/her prefrontal lobe and/or limbic system and committed a crime against another person, will he/or she go to hell? If that person is sent down to hell, why will God punish the object of his/her/its (nobody has identified the gender of God) creation? Despite the fact that behavior has various neuroanatomical, neurophysiological, and neurochemical bases, believers have a mindset that such hostile behavior is spiritually mediated and punishable by eternal suffering in hell. If there is a soul or spirit, how is it possible to burn something without chemical constituents. If there are organic and inorganic constituents in the soul, it is not possible to burn them endlessly because the chemicals will soon turn into non-inflammable materials, such as ash and dust, unless gasoline is poured into them.

For freethinkers, it is difficult to reconcile the value of religion, or even the existence of God, when wars among humans never end, considering that the system of life here on earth consists of the weak being killed by the strong, the strong being killed by the stronger, etc., and that the teachings of religion and the idea of God were obtained from various books and hearsay passed on from generation to generation by people with no knowledge of science and with unsophisticated intelligence compared with modern-day humans. This is the reality that believers find difficult to understand, and for them, everything converges into two words—God's will. ⊗

CHAPTER 20

Reflections, dilemma, and perspectives. Where do we go from here?

"Those to whom His word was revealed were alone in some remote place, like Moses. There wasn't anyone else around when Mohammed got the word either. Mormon Joseph Smith and Christian scientist, Mary Baker Eddy had exclusive audiences with God. We have to trust them as reporters—and you know how reporters are."
– *Andy Rooney (1919–2011), American radio and television writer*

"There is nothing more negative than the result of the critical study of the life of Jesus Christ. The Jesus of Nazareth who came forward publicly as the Messiah, who preached the Kingdom of God, who founded the Kingdom of Heaven upon earth, and died to give his work its final consecration, never had any existence."
– *Albert Schweitzer (1875-1965) – French physician, philosopher and humanitarian*

In the past, various secret societies and religions groomed Christianity and the story of Jesus Christ. Myths and rituals were supposedly recorded and then restructured and passed on from generation to generation. Those involved in the process had no knowledge of science, and the process of authenticating an event did not exist. They were powerful and influential and did everything to serve their ego. It is easily understandable that nonbelievers, freethinkers, and skeptics have wisely avoided the powerful grip of religion and its bandwagon of believers. They all challenge the existence of God, who, according to believers, "created" heaven and earth in just six days. The day and night difference between the God of the Old Testament (cruel, jealous, vindictive, genocidal, misogynistic, homophobic, and malevolent) and that of the New Testament (kind, merciful, and loving) is quite difficult to understand, yet believers claim that God does not change and is always the same, which is an egregious contradiction. Such is the power of the belief system in religion, especially when this system is inculcated into the central nervous system of the young and innocent.

Is religion bad or good? It is good because it promotes unity, but it is bad because it promotes bigotry and divisiveness—and there lies the contradiction. The belief in a powerful and controlling supernatural entitled to obedience, reverence, and worship, whose gender and physical form have not been identified, is quite difficult to understand, not to mention that this powerful being is invisible and exists in three forms! One form, Jesus Christ the son, was apparently visible and performed miracles, but other religions do not consider him as God or the Messiah. Another negative side of religion can be seen in the modus operandi of some televangelists who take advantage of the weak, especially those in need of help. Some conduct "healing sessions" to treat various medical disorders. On the stage, televangelists tap the forehead of the "sick" congregant, and in an instant, the follower falls to the floor and yells, "I am cured," hallelujah! One follower, who for years could not smell, was suddenly able to when the pastor moistened the nose with something (possibly ammonia). Suddenly, the woman claimed that her sense of smell came back! They earn non-taxable money, become millionaires and drive luxury cars. I must say, however, that some of them are genuinely and unequivocally pleasant, respectable, honorable, and honest.

Christians adhere to monotheism and insist that the three forms of God they believe in can be rolled into one. This sounds absurd, but believers call it the "mystery of faith." Christians and other religious groups ignore the existence of gods, those beings who descended from the sky and created mankind, reported by ancient civilizations in all parts of the world. They paved the way to understanding the world we live in, but the refusal to look back to the history of mankind dating back to way beyond the date of the creation of heaven and earth 6000 years ago is bothersome and clearly a hindrance to the advancement of science.

Modern science has taught us the mechanism of the creation of life through understanding the structure of DNA. We now understand why some people are born with genetic defects. Some of them die early, and some suffer for many years before dying of an incurable disease. Genetic engineering is here, albeit decelerated by religious issues. The mechanism of death of motor neurons in amyotrophic lateral sclerosis (or Lou Gehrig's disease) will soon be discovered, and appropriate treatment will be applied. Gene therapy may become routine when using embryonic cells. Nanotechnology may hold the key to the cure of cancer and other malignant disorders. All these will eventually happen without prayers.

A year ago, several members of my e-mail group became enraged when I brought up the idea that the Virgin Mary probably underwent in vitro fertilization and gave birth to Jesus Christ. Certainly, their naive reaction was understandable because of the teachings acquired during childhood that Mary was impregnated without sexual intercourse. How about asking yourself these questions. Was it disrespectful to Joseph that God (or the Holy Spirit) impregnated Mary?" How did she remain a virgin when, in fact, she was married to Joseph? What does the Bible say? In retrospect, it was conceivable that Mary was one of the many women involved in the creation experiments reported by ancient civilizations that involved the process of in vitro fertilization performed by gods from outer space to create humans in their own image. Some archeologists believe that the half-man and half-animal hybrids depicted in paintings on the walls of caves, in pyramids of the ancients, and in Greek mythology might have been representations of genetic manipulation and experimentation performed by advanced beings.

Sin is defined as anything against the moral principle and is punishable by eternal condemnation. If it is a crime, it is punishable by law. Some people are born with nice and pleasant personalities. Some are temperamental, and some are shy and introverted. Some who grow up without proper parenting may become criminals and substance abusers. Some have excessive hormonal activity and an abnormal limbic system; they may become rapists or pedophiles. Scientific studies have shown that differences in behavior are due to differences in the neuronal connections and activities of neurotransmitters in the brain. Such differences may drive a person to commit mass murders. Will these people go to hell? If they go to hell, why will God punish the object of his (her or its) creation? Can someday a person with abnormal neurons and neurochemicals in his/her limbic lobe be treated accordingly so that crimes against humanity can be prevented? At this time, and based on our current knowledge of neuroscience, sin and/or crime, as defined by religion and/or law, is punishable by eternal damnation and/or incarceration.

Nuclear weapons are proliferating all over the world. Is the total annihilation of humanity in the horizon? Do the frequent sightings of UFOs signify the efforts of advanced extraterrestrials who may have been monitoring humans for eons to prevent this from happening? Vitrification, a process that results in the formation of glass-like materials, usually results from nuclear explosions or volcanic activity. It is the result of extreme heat generated by nuclear blasts. They have been seen in India, Turkey, France, Scotland, and Ireland. Did a nuclear war happen before and send humanity back to the Stone Age?

The Sumerians and Mayans were established astronomers. Their writings revealed that they acquired knowledge of planets and stars from the sky gods. Some records pertaining to this knowledge might have been destroyed by wars, flood, and intentional destruction by colonizers proselytizing their religion. Knowledge about various star clusters, such as Pleiades, Sirius, and Orion, acquired by the ancients, is now being unraveled by modern astronomers. Humans will soon find ways to deflect incoming asteroids from colliding with the Earth. Unfortunately, humans will not be able to avoid the expanding sun, as it begins to lose its hydrogen fuel. It will expand and "swallow" the earth and the entire solar system. The only escape for humans is to move out of the earth and colonize other planets in another star system, the same fate that the sky gods that visited the planet earth may have met with, thousands or millions of years ago. Earthlings will be treated as "gods" when they encounter primitive beings in newly discovered habitable planets; the endless cycle of life will continue.

Is there something wrong living in a world without religion and/or belief in God? What is wrong with being aware of what is right and what is wrong, what is moral or immoral, or what is legal or illegal. Through the years, the success of indoctrinating children with religion and belief in an invisible being with an uncertain gender has been quite astounding but unfortunate. It can even be considered a form of parental malpractice and abuse, strictly speaking. As discussed in the previous chapters, evidence is compelling that indoctrination has neuroanatomical, neurochemical, and neurophysiologic bases. It is heartbreaking to imagine that instilling a child's brain with a belief system devoid of scientific reasoning and accepting a set of beliefs without any questions asked can be practically irreversible.

Historically, there is not a smidgen of doubt that millions of people died (and will continue to die) because of religion. Leaders of atheist countries have persecuted or executed people of various faiths to purge religion away from their countries. Atheists, agnostics, religious, and nonreligious people will continue to engage in heated debates. People will lose friends and will be ostracized when their ideology goes against the doctrines of believers.

I expect my friends, classmates, and relatives to adhere to their social and religious practices, and I will continue to be appalled, and sometimes amused (excuse my word), to read various stereotyped melodramatic remarks from my classmates and friends when a fellow classmate dies. Remarks like, "Now, he/she is now in the kingdom of the Lord," "He/She has now joined the Creator," "He/She is now living in eternal life" are just a few examples. One of my classmates used to post a blog titled, "Psalm a day," along with Bible verses, daily, for a year. For some reason, he quit doing it when very few classmates, and none later on, would bother to respond. Another classmate, after being offended by my negative blogs about religion, passionately vowed to continue teaching her children and grandchildren about God, Jesus Christ, etc. This is an unfortunate sequelae of a never-ending indoctrination process based on belief system. Sad but true.

In December 2021, a new survey conducted by the Pew Research Center showed that a group known as "no religious affiliation" now comprises 29% of Americans (up from 17% in 2009). The reasons for this are multifactorial and include ostracism, changes in attitudes toward homosexuality, rising discontent and distrust toward government officials, and changing demographics. The Pew survey was conducted among 3,937 respondents from May 29 to August 25, 2021. This number is expected to grow as science and technology advance.

The landscape of religion in the US has been changing quite remarkably. In a telephone interview conducted by the PEW Research Center in 2018 and 2019, 65% of American adults identified themselves as Christians, down by 12% over the past 10 years. Protestants and Catholics have experienced significant losses in numbers.

Why are Americans walking away from religion? Religion only has itself to blame. There is no science. There is nothing concrete and visible. The younger generation wants answers for everything. In today's technologically oriented society, there is always something that can be explained using the fingertips and the electronic rodent of the personal computer. They are beginning to realize that their problems cannot be solved by prayers. Mass and sermons are always predictably routine. Congregants kneel, sit, and stand in unison each time the priest proceeds to another segment of the mass. They sing the same songs and recite the same prayers again and again. There are so many religions in this multiethnic and multicultural society, members of which are covertly indifferent to each other. Broad-minded people have come to realize that the Bible's teachings are man-made and that religion and the 10 commandments are all about mind control. Some people do not see any correlation between belief in God and morality. The sight of a televangelist sobbing and crying, begging God to forgive him after sexual misconduct or fraudulent business practices, is quite appalling and disgusting. The old practice of condemning homosexuals and failure to recognize the scientific basis of certain sexual orientations is not acceptable to young generations. Interfaith marriages and mixed religious households are also significant factors. The intrusion of politics into religious practices is objectionable to some. Sexual misconduct by priests remains pervasive. One wonders whether priests who committed sexual misconduct have an innately abnormal neuronal network that dictates and predisposes them to commit such acts. If so, where is the accountability? If everything is genetically mediated and is someday explained and proven by science, the treatment may soon appear on the horizon. For now, an act that is against moral principles is a sin, and there is no other way to rename it at this time.

Predictably, it is quite obvious that an abrupt change in our belief system about religion can be shambolic, to say the least. For religious people, it is unimaginable to live without Christmas, Hanukkah, or Ramadan. For now and beyond, and for the unforeseeable future, there is no solution in sight. War, devastations from natural disasters, crimes, wide gaps between affluence and poverty, graft and corruption in the government, and famine in some parts of the world will perdure. Prayers will not prevent them from happening, and we know this for the reasons discussed in the previous chapters.

Believers will continue to use the Bible as the "authoritative" source of all creations in this vast and boundless universe. They ignore the dimension of a light year. They will continue to live inside a small room with only one narrow window and will remain imprisoned inside a steel cage of religion with no effort or desire to escape whatsoever. Children will continue to be indoctrinated by their parents resulting in their involuntary confinement inside a steel cage of religion.

The population in the Philippines, which is currently at least 100,000,000, will continue to grow . It cannot be controlled by abortion because abortion is a sin, according to the Catholic church. This is a galactic dilemma that Filipinos face. Combining this with a poor economy, the country remains behind other non-Catholic Asian countries. There appears to be no solution in sight at this time, as most Filipinos remain blindly adherent to the strict practice of Catholicism.

The discourse presented in this book's 20 chapters presents the major causal factors that forced me to deploy a valiant and daring effort to free myself from the powerful grip of religion. After reading several books on ancient civilizations and difficult-to-ignore evidences that their culture was likely influenced by advanced beings in the distant past, combined with gross inconsistencies and contradictions in the Bible and hypocrisies in religious practices, I find myself quite hesitant to offer an apology for telling my candid and honest opinion to religious people (including my friends, classmates, and relatives), who might someday, intentionally or unintentionally, read this book. I take pride in my honest views on religion, which are shared by freethinkers and broad-minded people from all walks of life. From the neurologic standpoint, I am simply lucky to harbor an innate and enduring neural fabric in my limbic lobe that is impervious to indoctrination enabling me to function usefully, and to develop ability, attitude and behavior of practical value in the society where I live without amalgamating them with religion and/or belief in supernatural.

To the readers of this book, believers and nonbelievers, try building another window in your tiny room, let the sunlight illuminate it, and be openly, pleasantly, and professionally percipient. Be realistic. Broaden your knowledge, expand your neural synapses, and keep your axoplasm flowing smoothly to keep your brain healthy.

"No matter how strongly held your belief might be or how significant they are to your identity as a human being, when presented with evidence that contradicts that which you hold dearly, it is your duty and obligation as a contributing member of society to discard your errant beliefs and attempt to realign your thoughts and actions with truth." – *Sam Harris*

Praise your Lord, Glory to your God in the Highest, and Peace be with you all (y'all) ଔ

BIBLIOGRAPHY

BOOKS (nonfiction)

Acharya, S, The Christ Conspiracy, The Greatest Story Ever Sold, Adventures Unlimited Press, 1999

Adams, F, Origins of Existence, The Free Press, 2002

Adams, RD, Victor, M, Principles of Neurology, McGraw-Hill, Inc., 1993

Alford, A, Gods of the New Millennium, Hodder & Stoughton, 1990

Allen, G, The Evolution of the Idea of God, The Book Tree, 2000

Basiago, AD, Thompson, EM, Fatima Revisited: The Apparition Phenomenon in UFOLOGY, Psychology, and Science, Anomalists Books, 2008

Baumgartner, A, Ye Gods!: A Dictionary of the Gods, Lyle Stuart, 1984

Bauval, R, Gilbert, A, The Orion Mystery, 1994

Berringer, R, Ancient Gods and their Mysteries, 2005

Blofeld, J, The Zen Teaching of Huang Po: On the Transmission of the Mind, Grove Press, 1994

Boulay, RA, Flying Serpents and Dragons: The Story of Mankind's Reptilian Past, Book Tree, 1999

Bushby, T, The Bible Fraud, The Pacific Blue Group, Inc., 2001

Childress, DH, Technology of the Gods: The Incredible Sciences of the Ancients, Adventures Unlimited Press, 2000

Childress, DH, Vimana Aircraft of Ancient India & Atlantis, Adventures Unlimited Press, 2001

Childress, DH, Lost Cities of Ancient Lemuria & The Pacific, Adventures Unlimited Press, 2002

Childress, DH, Extraterrestrial Archaeology, Adventures Unlimited Press, 2002

Cotterell, MM, The Supergods, Thorsons, An Imprint of Harper Collins, 1997

Cremo, MA, Thompson, RL, Forbidden Archeology: The Hidden History of the Human Race, Bhaktivedanta Book Publishing, 1998

Cremo, MA, Forbidden Archeology's Impact. Bhaktivedanta Book Publishing, 1998

Dawkins, R, The God Delusion, Houghton Mifflin Company, 2006

Downing, B, The Bible and Flying Saucers, Marlowe & Company, 1997

Dunn, C, The Giza Power Plant: Technologies of Ancient Egypt, Bear & Company 1998

Firestone, R, West, A, Warmick-Smith, S, The Cycle of Cosmic Catastrophes, Bear & Company, 2006

Francia, LH, A History of the Philippines, From Indios Bravos to Filipinos, Abrams Press, 2014

Friedman, ST, Flying Saucers and Science, Career Press, 2008

Gray, J, Dead Men's Secrets, TEACH Services, Inc. 2004

Greene, VM, The Six Thousand Year-old Space Suit, Maverick Publications, 1982

Hancock, G, The Sign and the Seal, A Touchstone Book, 1992

Hancock, G, Fingerprints of the Gods, Three Rivers Press, 1995

Hancock, G, Faiia, S, Heaven's Mirror: Quest for the Lost Civilization, Three Rivers Press, 1998

Hancock, G, Underworld: The Mysterious Origins of Civilizations, Three Rivers Press, 2002

Hancock, G, Supernatural: Meetings with the Ancient Teachers of Mankind, Three Rivers Press, 2006

Hapgood, C, Path of the Pole, Adventures Unlimited Press, 1999

Hapgood, C, Maps of the Ancient Sea Kings, Adventures Unlimited Press, 1996

Harris, Sam, The End of Faith: Religion, Terror and the Future of Reason, WW Norton, 2005

Hart, W, The Genesis Race: Our Extraterrestrial DNA and the True Origins of the Species, Bear & Company, 2003

Hausdorf, H, The Chinese Roswell, New Paradigm Books, 1994

Hitchens, C, God is not Great, Twelve Hachette Book Group USA, 2007

Horn, AD, Humanity's Extraterrestrial Origins, Silberschnur, 1994

Jones, AM, The Yahweh Encounters: Bible Astronauts, Ark Radiations, and Temple Electronics, The Sandbird Publishing Group, 1995

Jorde, LB, Carey, JC, Bamshad, MJ, Medical Genetics, Elsevier, 2019

Joseph, F, The Destruction of Atlantis: Compelling Evidence of the Sudden Fall of the Legendary Civilization, Bear & Company, 2002

Kenyon, JD, Forbidden History, Bear & Company, 2005

Kenyon, JD, Forbidden Religion, Bear & Company, 2006

Kenyon, JD, Forbidden Science, Bear & Company, 2008

Knight, C, Butler, A, Civilization One, Watkins Publishing, 2004

Kramer, PA, The Blood of Government: Race, Empire, the United States, & the Philippines, The University of North Carolina Press, 2006

Kramer, SN, History Begins at Sumer, University of Pennsylvania Press 1981

Kramer, SN, The Sumerians: Their History, Culture, and Character, University of Chicago Press, 2010

Leedom, TC, The Book Your Church Doesn't Want You to Read, Truth Seeker Books, 2003

Leon, D, Is Johovah an ET? Ozark Mountain Publishers, 2003

Lewels, J, The God Hypothesis, Wild Flower Press, 2005

Mills, D, Atheist Universe, Ulysses Press, 2006

Merritt, NJ, Jehovah Unmasked! Moon Temple Press, 2005

Moore, KL, Persaud, TVN, Torchia, MG, The Developing Human: Clinically Oriented Embryology, Saunders, 2015

Paine, T, Thomas Pain Collection: Common Sense, Rights of Man, Age of Reason, An Essay on Dream, Biblical Blasphemy, Examination of the Prophesies, Independently Published, 2019

Phelan, JL, The Hispanization of the Philippines, The University of Wisconsin Press, 1967

Pirsig, RM, Zen and the Art of Motorcycle Maintenance" An inquiry into Values, HarperTorch, 2006

Ridley, M, Genome, Perennial-An Imprint of Harper Collins Publishers, 1999

Rizal, J, Noli Me Tangere (Touch Me Not), Translated by Harold Augenbraum, Penguin Books, 2006

Rizal, J, El Filibusterismo, Translated by Harold Augenbraum, Penguin Books, 2011

Russell, B, Religion and Science, Oxford University Press, 1997

Sagan, C, Cosmos, Random House, 2002

Saranam, S, God Without Religion, The Pranayama Institute, 2005

Schellhorn, G, Cope, Extraterrestrials in Biblical Prophecy, Horus House Press, Inc.1989

Schievella PS, Hey! Is That You God? Sebastian Publishing Company, 1985

Schrag, P, Haze, X, The Suppressed History of America, Bear & Company, 2011

Sitchin, Z, Genesis Revisited, Avon Books, 1990

Sitchin, Z, The 12th Planet, Bear & Company, 1991

Sitchin, Z, When Time Began, Harper Collins Publishers, 1993

Sitchin, Z, Divine Encounters, Avon Books, 1995

Sitchin, Z. The Cosmic Code, Avon Books, 1998

Sitchin, Z, The Lost Book of Enki. Bear & Company, 2002

Sitchin, Z, Earth Chronicles Expeditions, Bear & Company, 2004

Steiger, B, Worlds Before Our Own, Anomalist Books, 2007

Steiger, B, Atlantis Rising, Galde Press, Inc., 2007

Stenger, VJ, GOD: The Failed Hypothesis, Prometheus Books, 2007

Tellinger, M, Slave Species of the Gods, Bear & Company, 2012

Temple, R, The Sirius Mystery: New Scientific Evidence of Alien Contact 5,000 Years Ago, Destiny Books, 1998

Temple, R, The Genius of China: 3000 Years of Science, Discovery and Invention, Prion, 2005

Thomas, RS, Intelligent Intervention: The Missing Link in the History of Human Evolution, Dog Ear Publishing, 2011

Turner, G, North Pole, South Pole, The Experiment, LLC, 2011

Von Daniken, E, Chariots of the Gods, A Berkley Book, 1968

Von Daniken, E, Gods from Outer Space, A Bantam Book, 1968

Von Daniken, E, The Eyes of the Sphinx: The Newest Evidence of Extraterrestrial Contact in Ancient Egypt, Berkley Books, 1996

Von Daniken, E, The Return of the Gods, Element Books, 1998

Von Daniken, E, Odyssey of the Gods, Element Books, 2000

Von Ward, P, We've Never Been Alone: A History of Extraterrestrial Intervention, Hampton Roads Publishing Company, Inc. 2011

Watson, JD, DNA: The Secret of Life, Alfred A. Knoff, 2004

ONLINE SOURCES

700,000-year-old Stone Tools Point to Mysterious Human Relative – www.nationalgeographic.com/science.article/stone

ANIMISM/Understanding Philippine Mythology – www.aswangproject.com/understand-philippine

Abu Al-Ala Al-Maari – en.wikipedia.org/wiki'Al-Maari

Before the Spaniards colonized the Philippines, the Filipinos were already civilized – putok.wordpress.com, /2010/11/23

The Laguna Copperplate Inscription: An Ancient Text That Changed the Perception of the History of the Philippines – www.ancient-origins.net/artifacts

Magellan expedition – en.wikipedia.org/Magellan

Did Christopher Columbus See UFOs? – Civilian Military – civilianmilitaryintelligencegroup.com

1492: Christopher Columbus, UFO's & the Bermuda Triangle – anomalyinfo.com/stories/1492- Christopher – Columbus

Christopher Columbus UFO Sighting in 1492 – gaf.news/2019/11/18 – Christopher Columbus - ufo

Lapu-Lapu – Wikipedia, en.wikipedia.org/Lapu-lapu

The Story of Lapu-Lapu: The Legendary Filipino Hero – theculturetrip. comp/asia/Philippines/articles

History of the Philippines – en.wikipedia.org/wiki/History of the Philippines

Philippines: The Spanish Period /Britannica – www.britannica.com/place/Philippines

University of Santo Tomas – Wikipedia, en.wikipedia.org/wiki/University of Santo Tomas

Abuses of the Spaniards towards Filipinos – prezi.com/3-3z6bm6uia/abuses-of-the-Spaniard

Catholic church sexual abuse cases by country – en.wikipedia.org/wiki/catholic church

Child sexual abuse in the institutional church, Philippines – www.manilatimes.net/2019/04/07

France Catholic Church abuse scandals – www.cnn.com/2021/10/05europe

Japan during World War II - Wikiquote – en.wikiquote.org/wiki/Japan_during_world_II

Americans in the Philippines – Wikipedia,
en.wikipedia.org/wiki/American

The Philippines, 1898-1946/US House of Representatives.
History.house.gov/Exhibition-and-Publication

Anti-Filipino sentiment – Wikipedia, en.wikipedia.org/wiki/Anti-Filipino

8 Dark Chapters of Filipino-American History We Rarely Talk About –
filipiknow.net/Philippine-american History

The Philippines Genocide up to 3 million Filipinos Killed –
medium.com/@BritsPhil/the -philippines-genocide

Mark Twain and the Anti-Imperialistic League/World History –
worldhistory.us/American-History/mark-twain

Japanese occupation of the Philippines – Wikipedia,
en.wikipedia.org/wiki/Japanese

Manila massacre – Wikipedia, en.wikipedia.org/wiki/manila

Michael Shermer - en.wikipedia.org/wiki/Michael_Shermer

Tiberius Julius Abdes Pantera –
en.wikipedia.org/wiki/Tiberius-Julius-Abdes-Pantera

The "Jesus son of Pantera" Traditions –
jamestabor.com/the –Jesus-son-of-pantera-traditions

History's most famous artworks "littered" with aliens and UFO's –
www.sun.co.uk/the/9446848/famous-artworks

10 Bible Accounts That Could Be Interpreted as UFOs or Aliens –
listverse.com/2017/07/10-bible-accounts

We're closing in on them: UFO expert on Pentagon findings –
www.newsnationnow.com/on-balance-with-leland

10 deadliest natural disasters in the Philippines –
newsinfo.inquirer.net/524569

World War II Casualties – Wikipedia-
en.wikipedia.org/wiki/world_war_II casualties

50 Stephen Hawking Quotes on Life, Religion, and the Universe –
everydaypower.com/Stephen-hawking-quote-2

Albert Einstein: Quotes on God, Religion, Theology –
spaceandmotion.com/albert-einstein-god-religion

Anti-Christian Quotes – www.goodreads.com/quote/tag/anti-christian

What's your religion? In U.S., a common reply now is "None," –
news.yahoo.com/what's-religion—us-common-reply

Pew Research Center – Wikipedia,
en.wikipedia.org/wiki/Pew_Research Center

Biblical Contradictions – www.atheists.org

Exodus: Why Americans are leaving religion—and why they're unlikely to come back – www.prri.org

In U.S., Decline of Christianity continues to rapid pace – www.pewforum.org

Pew Research: Why young people are leaving Christianity – answersingenesis.org/church/pew-research-why

Catholicism and the Philippine Population Problem – www.jstor.org

Is Catholic Church's influence in Philippines fading? – BBC News – www.bbc.com/news/world-asia 27537943

Thomas Paine's quotes on religion – www.learnreligion.com/top_thomas_pain_quotes

Robert Ingersoll's quotes about religion – www.azquotes.com/author/7171

Albert Einstein Quotes on God, Religion, Theology – spaceandmotion.com/albert-einstein-god-religion

Robert M. Persig – Freedom From Religion Foundation – ffrf.org/ftod –cr/itm/14883-robert-m-persig

Your Favorite Anti-religious quotes/Rational Response Squad – www.rationalresponders.com/atheist

42 Funny Atheist Quotes About God, Existence, Fate and Life – www.geckoandfly.com/23894/atheist-religious

The Forged Origins of the New Testament by Tony Bushby – exminister.org/Bushby-forged-origins-New Testament

Pope Leo X – en.wikipedia.org/wiki/Pope_Leo X

American religious group vary widely in their view of abortion – www.pew research.org/fact

Jehovah's Witness Ethics – religionfacts.com/jehovahs-witness/ethics

Religion and Science – The New York Times – www.nytimes.com/1930/11/09

Christian Atrocities – stellarhousepublishing.com/victims

Spanish Inquisition – en.wikipedia.org/wiki/Spanish_Inquisition

Cloning Fact Sheet-Genome.gov – www.genome.gov

Gene Therapy News – www.sciencedaily.com

MEDICAL, NEUROLOGICAL and NEUROSURGICAL JOURNAL ARTICLES

Adolphs R, Tranel D, Damasio H, Damasio A. (1994) Impaired recognition of emotion in facial expressions following bilateral damage to the human amygdala. Nature. 372 (6507):669-672

Agorastos, A, Demiralay, C, Huber, CG. (2014) Influence of religious aspect and personal beliefs on psychological behavior: focus on anxiety disorder. Psychol Res Behav Manag. 7:93-101

Anderson PD, Bokor G. (2012) Forensic aspects of drug-induced violence. J Pharm Pract. 25:41-49

Benson, DF. (1991) The Geschwind Syndrome. Adv Neuro.l 55:411-421

Bidzan, L, Bidzan, M, Pachalska, M. (2012) Aggressive and impulsive behavior in Alzheimer disease and progression of dementia. Med Sci Monit. 18:CR 182-CR 189

Blanke, O. (2004) Out of body experiences and their neural basis: They are linked to multisensory and cognitive processing in the brain. BMJ. Dec 18; 329 (7480):1414-1415.

Blanke, O, Mohr, C. (2005) Out-of-body experience, heautoscopy, and autoscopic hallucination of neurological origin: Implications for neurocognitive mechanisms of corporeal awareness and self-consciousness. Brain Res Rev. 50:184-199

Brewerton, T. (1994) Hyperreligiosity in Psychotic Disorders. J Nerv Ment Dis. 182:302-304

Cardinal RN, Parkinson JA, Hall J, Everitt BJ. (2002) Emotion and motivation: the role of the amygdala, ventral striatum, and prefrontal cortex. Neurosci Biobehav Rev. 26:321-352

Coupland, CAC, Hill, T, Dening, MD. (2019) Anticholinergic Drug Exposure and the Risk of Dementia: A Nested Case-Control Study. JAMA Intern Med. 179:1084-1093

Davis M, Whalen PJ. (2001) The amygdala: vigilance and emotion. Mol Psychiatry. 6:13-34

Del Arco A, Mora F. (2009) Neurotransmitters and prefrontal cortex-limbic system interactions: implications for plasticity and psychiatric disorders. J Neural Trans (Vienna). 116:941-952

Fields, CM, Marcuse, LV. (2015) Palinacousis. Handb Clin Neurol. 129:457-467

Goetz, CG, Tanner, CM, Klawans, HL. (1982) Pharmacology of hallucinations by long-term drug therapy. Am J Psychiatry. 139:494-497

Greenberg, D, Huppert JD. (2010) Scrupulosity: a unique subtype of obsessive compulsive disorder. Curr Psychiatry Rep. 12(4):282-285

Heimer, L, Van Hoesen, GW. (2006) The limbic lobe and its output channels: implications for emotional functions and adaptive behavior. Neurosci Biobehav Rev. 30:126-147

Heydrich, L, Lopez, C, Seeck, M, Blanke, O. (2011) Partial and full own-body illusions of epileptic origin in a child with right temporoparietal epilepsy. Epilepsy & Behav. 20:583-586

Jacobs, L, Feldman M, Diamond, SP, Bender, MB. (1973) Palinacousis or Recurring Auditory Sensations. Cortex. 9:275-287

Khoubila, A, Kadri, N. (2010) Religious obsession and religiosity. Can J Psychiatry. 55(7):458-463

Kringelbach, ML (2005) The human orbitofrontal cortex: linking reward to hedonic experience. Nat Rev Neurosci. 6:691-702

Kucker S, Tollner K, Piechotta M, Gernert M. (2010) Kindling as a model of temporal lobe epilepsy induces bilateral changes in spontaneous striatal activity. Neurobiol Dis. 37:661-672

Kuhn, D, Greiner, D, Arseneau, L. (1998) Addressing hypersexuality in Alzheimer Disease. J Genrontol Nurs. 24:44-50

Lane, SD, Kiome, KL, Moeller FG. (2011) Neuropsychiatry of Aggression. Neurol Clin. 29:49-64

Lazaro, RP (1983) Palinopsia: Rare but ominous symptoms of cerebral dysfunction. Neurosurgery. 13:310-313

Li, X, Song, R, Qi, X, Xu, H, Yang, W, Kivipelto, M, Bennett, DA, Xu, W (2021) Influence of cognitive reserve on cognitive trajectories. Neurology. 97:e1695-e1706

Levitan, L, LaBerge, S, DeGracia, DJ, Zimbardo, PG. (1999) "Out-of-body experiences," dreams, and REM sleep. Sleep and Hypnosis. 1,186-196

Mayo, T. (2014) Brain-dead and pregnant in Texas. Am J Bioeth. 14:15-18

McNamara, JO. (1986) Kindling model of epilepsy. Adv Neurol. 44:303-318

Najar, J, Ostling, S, Gudmundsson, P, Sundh, V, Johansson, L, Kern, S, Guo, X, Hallstrom, T, Skoog, I. (2019) Cognitive and physical activity and dementia. Neurology. 92:e1322-e1330

Nelson, KR, Lee SA, Schmitt, FA. (2006) Does the arousal system contribute to near death experience? Neurology. 66:103-100

Nelson, K. (2015) Near-Death Experiences: Neuroscience Perspectives on Near-Death Experiences. Mo Med. 112:92-06

Nichols, DE. (2004) Hallucinogens: Pharmacol Ther. 101:131-181

Pope, T. (2018) Brain Death and the Law: Hard Cases and Legal Challenges. Hastings Center Report. 48:546-548

Perani, D, Farsad,M, Ballarini, T. Lubian,F, Malpetti, M, Frachetti, A, Magnani, G, March, A. Abutalabi, J (2017) The Impact of Bilingualism on brain reserve and metabolic connectivity in Alzheimer's dementia. PNAS.114:1690-1695

Russell, JA, Epstein, LG, Greer, DM, Kirschen, M, Rubin. MA, Lewis, A. (2019) Brain death, the determination of brain death, and member guidance for brain death accommodation request. Neurology. 92:228-232

Sala, A, Malpetti, M, Farsad, M, Lubian, F, Magnani, G, Polara, GF, Epiney, JB, Abutalebi, J, Assal, F, Garibotto, V, Perani, D. (2021, Nov 2. Doi:10.1002.hbm.25605

Servidei, S, Lazaro, RP, Bonilla, E, Barron, KD, Zeviani, M, DiMauro, S. (1987) Mitochondrial encephalomyopathy and partial cytochrome c oxidase deficiency. Neurology. 37:58-63

Slachevsky, A, Pena, M, Perez, C, Bravo, E, Alegria, P. (2006) Neuroanatomical basis of behavioral disturbances in patients with prefrontal lesions. Biol Res. 39:237-250

Stern, Y. (2012) Cognitive reserve in ageing and Alzheimer's disease. Lancet Neurol. 11:1006-1012

Wei, W, Chinnery, PF. (2020) Inheritance of mitochondrial DNA in humans: implications for rare and common diseases. J Intern Med. 287:634-644

Wijdicks, EFM, Varelas, PN, Gronseth, GS, Greer, DM. (2010) Evidence update: Determining brain death in adults. Reports of the Quality Standards Subcommittee of the American Academy of Neurology. Neurology. 74:1911-1918

PERIODICALS

Farkas V. Mysterious Vitrifications Around the World. Legendary Times. 2006, 7:47-48

Fiebag P, Eenboom A, Belting P. The Theotechnology of Ezekiel. Legendary Times. 2005, 7:18-21

Kluger J. What Makes Us Good/Evil. Time Magazine. Dec 3, 2007, 53-60

Langbein WJ. The Spaceships of Ezekiel. Legendary Times. 2006, 7:39-41

Sasoon GT. The Ancient of Days: Deity or Manna-machine. Legendary Times. 2000, 2:4-7

Von Daniken, E. Stonehenge - Message From the Stars? Legendary Times.2002. 4:13-19

Von Daniken, E. Of Monsters & Man-Animals. Legendary Times. 2003, 5:9

Von Daniken, E, Wishful Egyptologic Thinking. Legendary Times. 2002. 4:7-15

Winfield, N (The Associated Press). Indigenous tell pope of abuses endured at Canada Schools. Schenectady Gazette, March 29, 2022

ABOUT THE AUTHOR

Dr. Rey Lazaro resides in Guilderland, a small quiet town in Upstate, New York, west of Albany. A graduate of University of Santo Tomas Medical School in Manila, Philippines in 1970, he received his neurological training at St. Vincent Hospital and Medical Center of New York, and neuromuscular disease training at Vanderbilt University Medical Center in Nashville, Tennessee. He is board-certified in Adult Neurology and Electrodiagnostic Medicine. Most of his time in semi-retirement is devoted to clinical research and literature search on topics such as unexplained ancient mysteries, religion and science of ancient civilization, astronomical and geological phenomena, and personality profiles of deeply religious people.

Lightning Source UK Ltd.
Milton Keynes UK
UKHW010658090223
416681UK00007B/1824